MAPS THAT MADE HISTORY

MAPS THAT MADE HISTORY

The influential, the eccentric and the sublime

LEZ SMART

the national archives

To my family: Jenny, Joseph, Holly and Harry

First published in 2004 by

The National Archives
Kew, Richmond, Surrey
TW9 4DU, UK

vww.nationalarchives.gov.uk/

The National Archives was formed when the Public Record Office and Historical Manuscripts Commission combined in April 2003.

A catalogue record for this book is available from the British Library.

ISBN 1 903365 64 3

Jacket illustration: a chart of the British Isles from John Seller's *Atlas Maritimus*, first published in *c.* 1678. John Seller (*c.* 1630–1697) was the first Englishman, in England, to compete with the Dutch as a producer of world marine atlases. Although as a nonconformist he was found guilty of conspiring to kill King Charles II, he was reprieved, and became Royal Hydrographer in 1671. He ran a nautical chart business from a shop near the Tower of London, and was also an instrument and globe maker.

Frontispiece: 'Departure from Lisbon', by Theodore de Bry (see pp. 60–65).
Page 8: *Twilight in a Forest*, by Adrian Scott Stokes (see pp. 156–61).
Page 186: Poster for the London Underground, *c.* 1930s (see pp. 112–17).

Editorial, design and production by The Book Group, Somerset
Printed in Slovenia by MKT PRINT

Contents

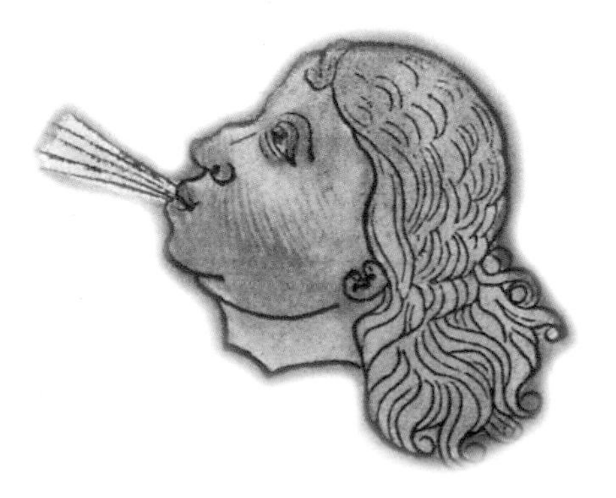

Preface

I'm not absolutely sure when I first became aware that I found maps fascinating. I think it was probably on family holidays when I was a child. Each year our family made its way across Leicestershire and Lincolnshire in a small car to the east coast holiday town of Mablethorpe. As the eldest child I was charged with the responsibility of reading the map and advising my Dad as to the correct route. This invariably took the form of identifying the next place we should come to, calling it out as we reached it and then identifying the next one. Most of these places were tiny villages with wonderful names such as Frisby on the Wreake, Waltham on the Wolds, Tattershall and everyone's favourite, Maltby le Marsh – the reason for the popularity of the latter arising from the fact that it meant we were 'almost there'! I remember wondering who had given these places their names? Why did Maltby have a French name? Why had they made the streets so 'wiggly' (very much a Leicester word)? I didn't just want to use the map, I wanted to know more about it, the places on it and the people who had made it the way it was. Although I didn't realize it at the time the asking of these questions was the beginning of a lifelong fascination. As my bookshelves and the walls of my house bear witness, the writing of this book is just one manifestation of this interest.

In the process of writing it I have examined maps in the hushed Map Rooms of the National Archives and the British Library – maps and locations far removed from the crumpled road atlas of the East Midlands flattened out on the dashboard of a Ford Popular. And yet the questions asked of these often unique and priceless maps was not that dissimilar from my questions as a child. Who made this map? Why did they make it? When did they make it? Why did they decide to give that name to that place? What does that line/symbol mean? How did they know it looked like that?

These, and further questions, are the ones asked in the following pages. As will be seen, sometimes the answers can be ascertained, sometimes they have to be surmised and sometimes we just don't know. And that's why they continue to fascinate.

Lez Smart

Acknowledgements

Without the persistence and tenacity of Sheila Knight, Deputy Publishing Manager at the National Archives, this book would not have been written. I'd like to thank her for seeing the potential behind the ideas we discussed at out first meeting and being supportive all the way through.

Thanks are due to Paul Johnson and especially Hugh Alexander who form the National Archives Image Library team. Their suggestions and advice were invaluable. Thanks also to Jenny Speller whose painstaking work in tracking down 'permissions to use' formed one of those vital but unsung tasks. Thanks to all at The Book Group for their editorial and design contributions.

However, my major thanks must go to my wife Jenny. Her unflagging support for this project and her willingness to listen to my enthusiastic account of yet another obscure detail from an equally obscure map was beyond the call of duty. Thanks Jen.

Finally, a word of thanks to Harvey for taking up his position under the desk each day with never a complaint.

La tour de babilon
Nembroth

Workmen use a wattled ladder to build the Tower of Babel that would reach from earth to heaven. Its location in the Bible allowed it to be plotted on to maps *(see pp.174–9).*

Introduction

We need maps. We have always needed maps. We always will. The marks found in the caves of our ancestors have been interpreted as maps and every society that has existed throughout history, in every corner of the globe, has created its own maps. They form one of the ways by which each society has sought to represent, record and communicate its world.

While the scale, size and detail may vary all maps show the relationship between one place and another. They include the details that are, or were, deemed significant and omit the ones that are or were not. No map has ever, nor can ever show 'everything'.

When faced with a map most people will immediately seek to identify a refer ence point from which to start: 'There's our street'; 'I went there for my holiday'; 'Your grandmother was born there' or even 'One day I'm going to go there'. This is perfectly natural because it is essential to get one's bearings and establish a point from which to make sense of the information on the map.

This ability to 'read' a map is part of a skill known as visual literacy and involves being able to make sense of symbols, images and colours either alongside or in the absence of words. This literacy was present in the earliest humans, and has continued through the ancient civilizations into modern times. It is no less important a skill to possess than reading and writing. As the evidence from the earliest cave markings to the maps drawn by children showing their way to school illustrate, the ability to make and use maps may be a natural skill rather than a learned one.

Every map has been created at a particular time and place. Most have been created for an intended purpose and often for an intended audience. As such, each map can provide us with an insight into a particular period in history, the people who

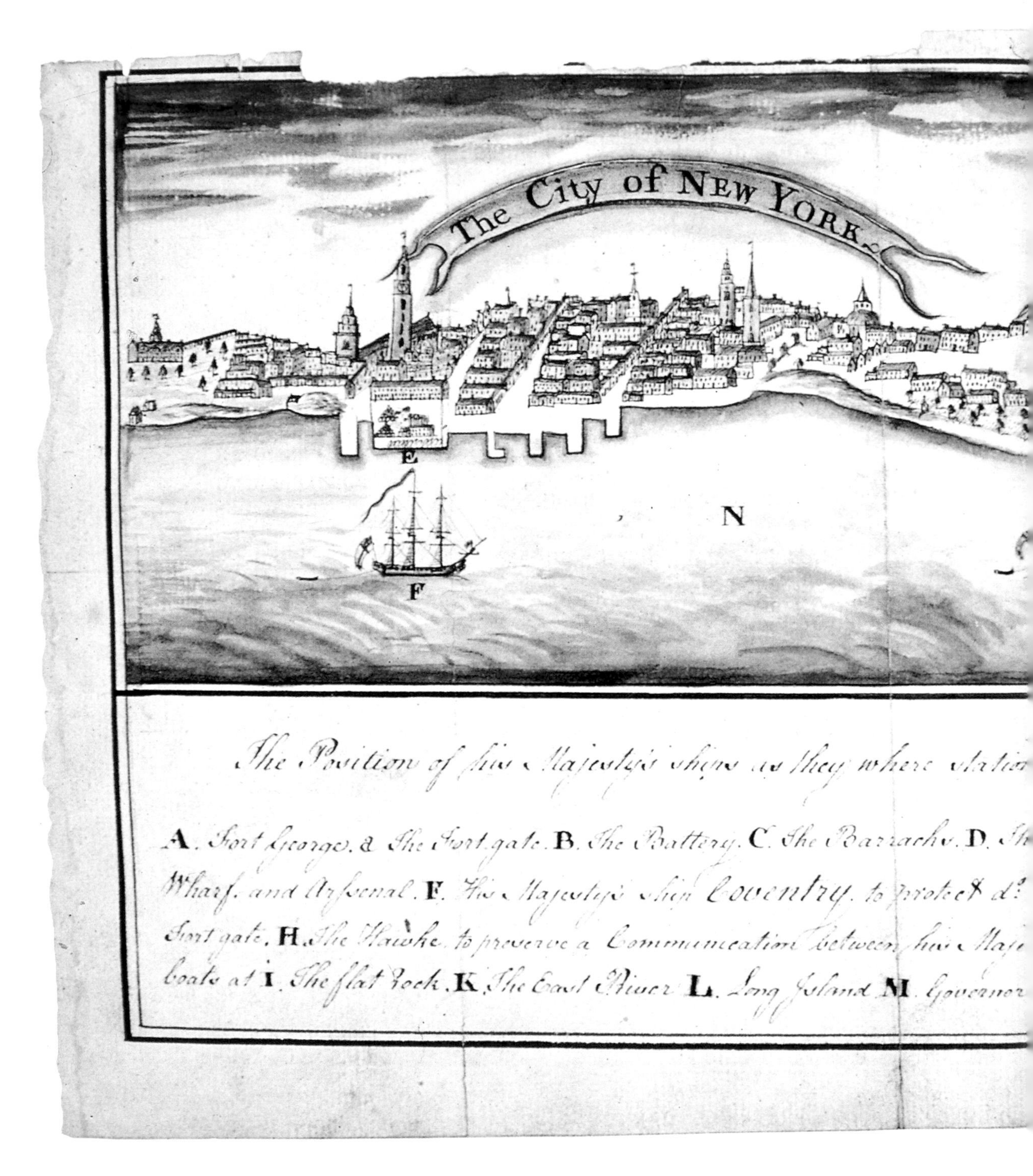

lived there and what was significant to them. But like all artefacts that have survived from earlier periods maps do not speak for themselves: they need to be explored and interpreted. This is what this book seeks to do.

Each map has its own story to tell. As the following pages reveal sometimes this story is about the events the map records, sometimes it revolves around how it was used, while at other times the cartographer himself is central to the story. In many

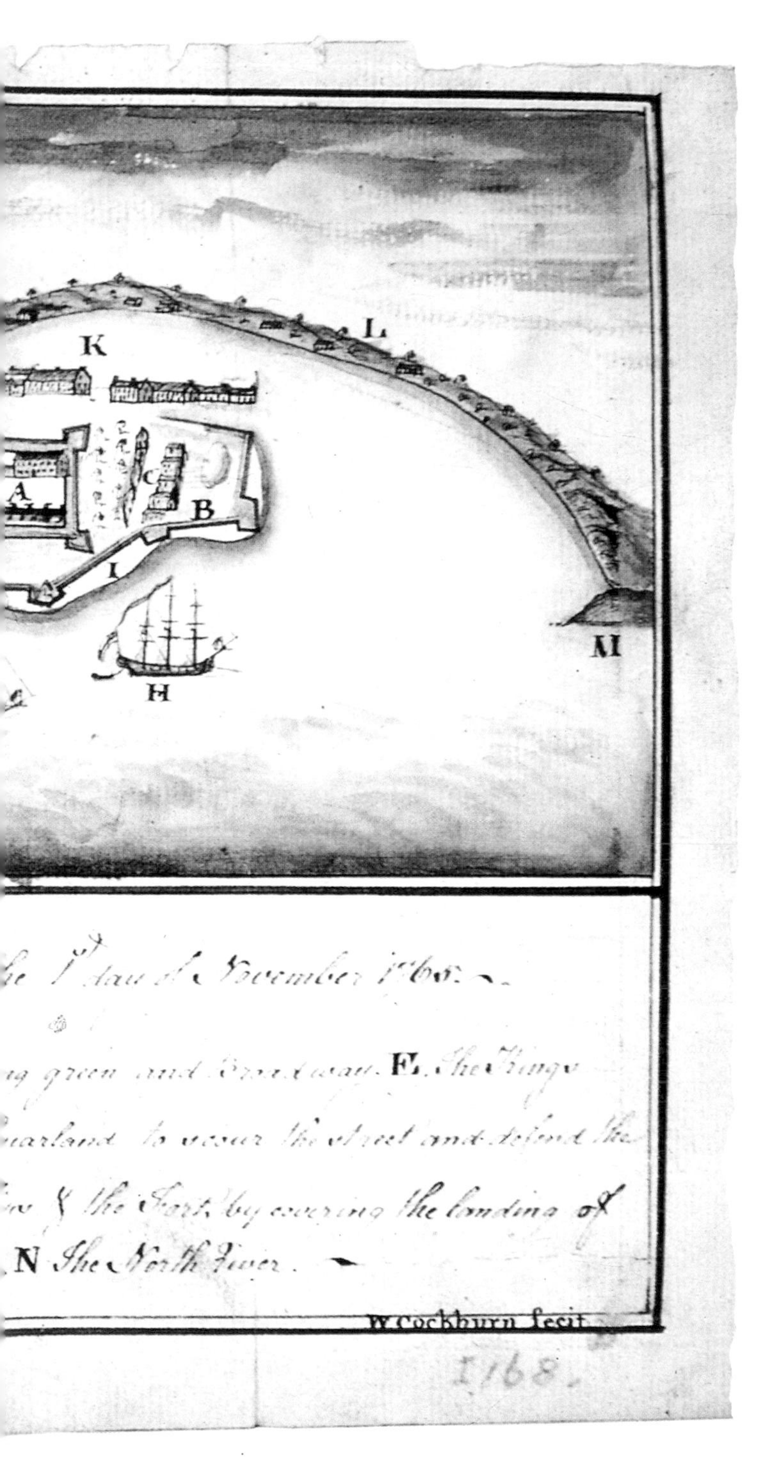

Map of New York, by W. Cockburn, 1767. The map shows the position of Royal Navy ships on 1 November 1765, deployed in connection with the riots against the Stamp Act; this unpopular act was ultimately the fuse that ignited the American War of Independence *(see pp. 88–93).*

cases it is a combination of these different elements. The story may involve kings and queens, or monks, generals and sea captains, or it may describe the actions of ordinary soldiers or Londoners going about their daily business. History itself is like that and maps contribute to our understanding of the past.

When we look at a map today we usually start from an assumption that it is accurate. If a place is marked then it must exist and if we so chose, we could visit it. If we were to do so and find that it wasn't there, surprise and probably indignation would be our response.

However, a cartographer at any point in history can only include the information that is available, which may be inaccurate or incomplete. As several of the maps in this volume show this has led to the creation of maps featuring some fantastic locations, events and characters. It would be totally inappropriate to dismiss these as simply 'incorrect' or even 'silly', whether it is the existence of a gigantic southern continent, the exact location of Noah's Ark or the presence of giants and monsters. It would be wrong to smile patronizingly at the maps that show California as an island, or a Great River flowing through the centre of Australia or the location of El Dorado. Like written accounts of the past all maps are 'prisoners of their time' but they allow us to see how the world or a particular part of it was viewed and understood by the mapmaker and his audience at that time. This last point is important for most of the maps in this collection are of European origin and this European perspective has to be appreciated. Captain Cook's 'voyage of discovery' along the eastern coast of Australia would not have impressed the Aborigines whose families had been living there for thousands of years!

Maps were often embellished with extra details and a cartouche, such as this fine example from the map of the Battle of Culloden (see pp. 122–7).

Numerous books have been published on the subject of maps. Many deal with collecting, some with the mapmaker's art, while others focus on the historical development of cartography itself. Very few adopt the approach taken here and explore individual maps as a means of looking into the history of the period in which they were created. In effect, to use them as a 'window' into times now passed. The attempt to do this has informed the way they are presented in the following pages. All the maps have been reproduced as large as possible to enable the details to be seen easily, because it is often the detail that is crucial to the story.

As I assembled the histories behind the maps my magnifying glass was a vital tool and always at hand. The sections of each map that have been 'extracted' and expanded for a more thorough exploration are simply what readers would find if they were to use a magnifying glass themselves. It is often surprising what can be missed with the naked eye.

Finally, it needs to be remembered that what was placed on the map was considered significant and what omitted not so. As will be revealed, 'the medium' was as much part of 'the message' in the past as it is today and the use of subliminal techniques to get this across clearly predate the work of the modern advertisement.

All history is a process of interpretation and this applies here. The choice of the maps themselves, the highlighted extracts and the people and events that are mentioned have all been subjected to this process. Other interpretations and the accordance of significance to particular fragments are certainly possible. Indeed, they are invited.

On the morning I commenced writing this Introduction the radio news announced that the Royal Geographical Society in London was opening its archives for use and that this included some 2,000,000 maps, many of whose existence was not previously known! An internet search using the word 'maps' on the Google site brings up a staggering 125,000,000 responses. The maps we now use will be scrutinized by future generations as they seek to deepen their understanding of our twenty-first-century world. They too will attempt to establish 'the story behind the map' and, as is the case here, also find there is often more to this than initially meets the eye.

The 25 maps presented here are clearly only a miniscule fraction of the ones that survive. It is hoped that the insights they provide will prove both as illuminating and fascinating to the reader as I found during the process of putting them together.

The Romans were prolific road builders creating a vast network throughout their empire. This bas-relief from the first century AD demonstrates the typical construction process with a surface of compact cobbled stones over a layer of cement (see pp. 26–31).

Chapter One

The Early Mapmakers

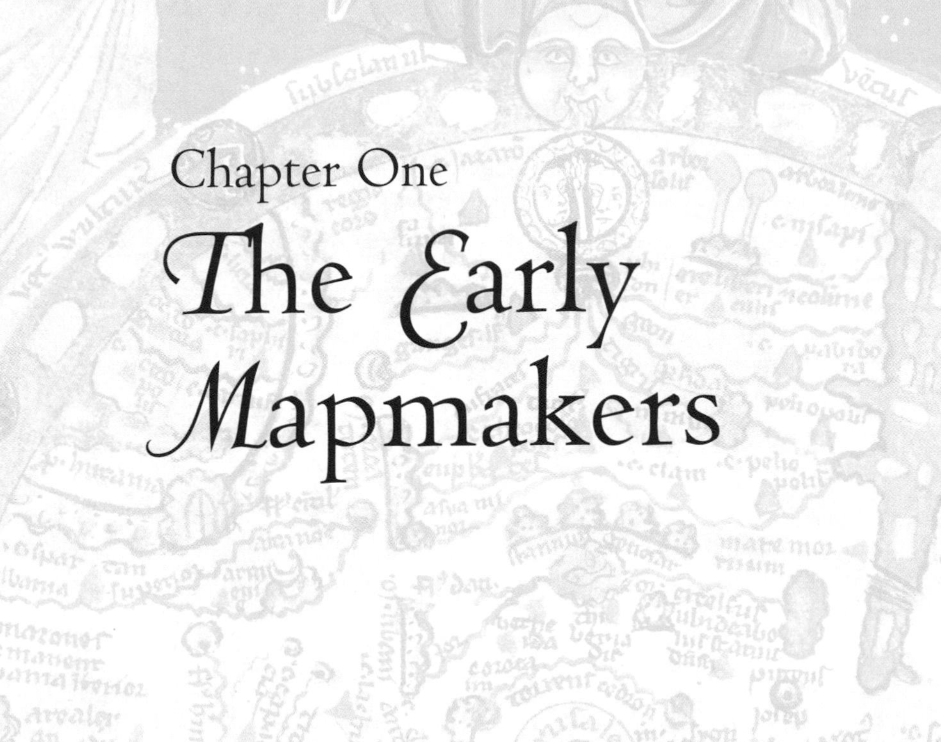

The desire to exercise a degree of control over one's environment is a distinguishing human feature. One of the prerequisites of being able to do this successfully is to have an awareness of where one is and what it's like in relation to other places. Our understanding of our world has always been different in different places and at different times because it has, and always will be, a culturally constructed understanding. The maps chosen for inclusion in this chapter allow us to see how the world, on both a large and small scale, was constructed in ages gone by.

A scribe working in one of the many religious houses in Britain during the twelfth century.

What survives from earlier periods of history is determined more by chance than intention. The natural chemical processes of decomposition combined with the destructive human elements of neglect and conflict have always, and will continue to take their toll. The fragile nature of paper and parchment on which maps have been traditionally drawn has made them extremely vulnerable. The ones that feature in this chapter are survivors. Most of their contemporaries have been destroyed and we are fortunate that these did not suffer the same fate. Their continued existence, however, is by no means guaranteed and while great care is now taken in their storage and display, their long-term survival is also likely to involve an element of good fortune.

What needs to be remembered as one considers the maps in this chapter is that they were created (with the exception of the copy of the Ptolemy map) before the invention of the printing press in 1450. Each was painstakingly hand drawn. Today we 'make a copy' of something without a second thought. In the pre-printing era this meant someone had to literally copy the map they had on the table in front of them. It was a skilful and time-consuming practice and the majority of this work was undertaken in monasteries, abbeys and other religious houses. All five maps have a direct link with such settings. This connection acts as a reminder that whatever views we may hold with regard to Christianity the role of religious houses in the collection and transmission of accumulated knowledge over the centuries has been highly significant. In two cases, Nicolaus Germanus (pp. 20–25) and Matthew Paris (pp. 38–43), we know the names of those involved and acknowledge their contribution. But the names of the monks who copied the epic Peutinger Table (pp. 26–31) or the beautiful Psalter (pp. 32–7), or who drew the stylized Chertsey Abbey sketch (pp. 44–9) are lost in the mists of time.

As ever, there are exceptions to this rule and at times religion has led to the systematic destruction of books and manuscripts and even to the death of those who created them. One of the maps makes this point quite poignantly. The absence of an original Ptolemy map is very likely to be due to the act of vandalism inspired by religious fervour that destroyed what was at the time the largest library in the world at Alexandria. History, as ever, gives contradictory messages.

Each of the following maps, in its idiosyncratic way, provides us with valuable information about the time in which it was created and the period it represents. However, as the stories behind the Peutinger and Ptolemy maps reveal, these two cannot be assumed to be the same.

The variation in style and content between the maps included is obvious as they are perused. The incredible number of mathematical calculations involved in the creation of the Ptolemy map contrasts sharply with the map of Chertsey Abbey with its simple diagrammatic representations. Whether one is more important than the other is an interesting question to pose and would give rise to lively debates in 'learned circles'.

In the Introduction it was noted that no map can ever show 'everything' and that what is included and what is omitted from a map is part of the story itself. Although the mapmaker's decision-making would be determined by the map's purpose, he was also part of the society he sought to represent, and his work cannot but reflect this. Just as Matthew Paris's 'pilgrimage' map of Britain enables us to see how one of the leading scholars of his day perceived the physical link between England and Scotland so it also allows us to deduce something about the state of the road system, or lack of it, at this time. The Psalter map on the other hand was not intended to help anyone get to anywhere – except perhaps to Heaven – but its combination of geographical features with incidents from the Bible allows us to attempt to deduce a little about the state of the Mediterranean world's knowledge of Africa and the Middle East.

CAVRVS·CHORVS·VEL·IAPIX·SIVE·ARGESTES
CIRCTVS VEL·TRESIAS
SEPTENTRI VEL·APARCTIAS
FAVONIVS·ZEPHIRVS
AFRICVS·VEL·LIBS
LIBONOTVS·EVROAVSTER
AVSTER·VEL·NOTVS
mare glaciale
EVROPA
Pontus euxinus
LIBIA·INTERIOR
AFFRICA
Circulus equinoctialis
ETHIOPIA·INTERIOR
ARABIA·FELIX
MARE·INI
Sinus Barbaricus
MAR
Tropicus Cancri

The Ancient World Recreated

Map According to Ptolemy's Projection

By Nicolaus Germanus, 1482

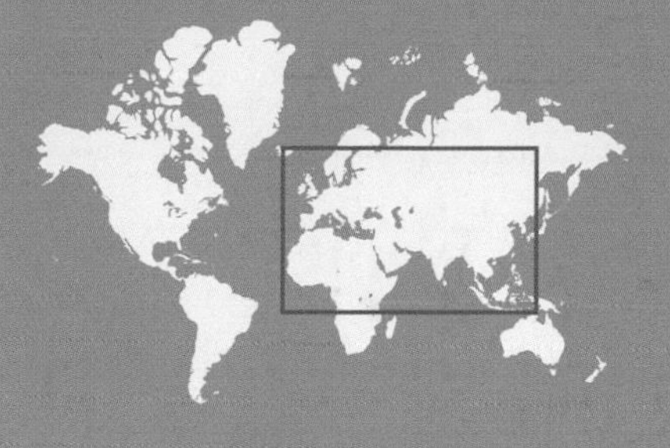

The Ancient World Recreated
Map According to Ptolemy's Projection

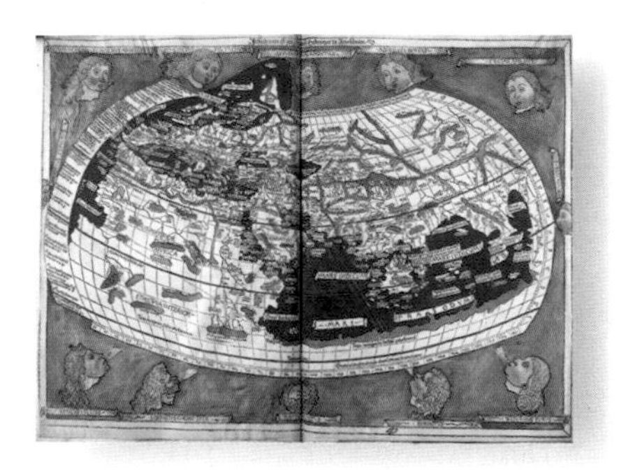

Some maps have a date on them. Others can be dated by identifying the creator and working out when he (and it usually was a he) lived. It is important to be able to establish the date when a map was created because it gives an insight into the world as known and perceived at that time. Sometimes one finds experts don't agree on the exact date but will agree it was between two dates during a specific time period. This wonderfully coloured map of the world defies this approach. In some works it is dated 1482 while in others 150! The explanation as to how such a situation can exist reveals a great deal about the history and development of cartography. Claudius Ptolemaeus, known as Ptolemy, lived between *c.* AD 100 and AD 168 and was one of the leading scholars of his time. He lived and worked in Alexandria, the Egyptian city that was the intellectual centre of the western world with the most extensive library that had ever been created.

Ptolemy brought together his mathematical and astronomical knowledge and skills and applied them to mapmaking. He had the mathematician's fascination with precision and set out to show the relationship of one place to another not figuratively but accurately. Based on a calculation of the world's circumference of 18,000 miles, he further developed the grid system of latitude and longitude devised by Marinus of Tyre. While some of the details on the map may be a little strange to the eye the lines of latitude running parallel to the equator crisscrossed by the lines of longitude running north–south in graceful arcs are familiar to anyone who has ever opened an atlas. Within this framework Ptolemy was able to establish coordinates and in his major work *Geographia* he listed over 8,000 places and their respective coordinates. These were given in the degree, minute, second division we use today. For Ptolemy it was a mathematical exercise and we will never know whether he actually drew any maps from these. If he did they were lost, possibly when the famous Library of Alexandria was burnt down by fanatical Christians in AD 390 – an early example of the conflict between faith and science. But at least one copy, maybe more, had been made of Ptolemy's works and these survived in Byzantium. For the next 1,000 years his writings were used and developed by Arab scholars while Europe remained in ignorance of his legacy. It was not until the emergence of the Renaissance in Italy and its fascination with the classical world that Ptolemy's *Geographia* was translated into Latin and his ideas became accessible to European scholars once again.

However, there were no maps in the surviving version, simply the instructions and advice on mapmaking and the lists of coordinates. Working with these 1,200-year-old coordinates, early Renaissance scholars, such as the Benedictine monk Nicolaus Germanus, slowly assembled these references into a map of Ptolemy's world.

Portrait of Ptolemy, astronomer, geographer and mathematician, c. *1476 by Justus van Gent.*

So what date should this map carry? Should it be 1482, for this is when it was drawn by Nicolaus? Or should it be a copy of the Ptolemy map *c.* AD 150, even though Nicolaus was not actually copying anything and no one knows for certain whether Ptolemy himself ever drew a map from his data. Whatever, Ptolemy's scientific approach was the first known projection of a sphere on to a plane, and would influence all future attempts to present the information about a spherical world on a flat map.

In the late fifteenth century it had a more dramatic and immediate impact. It was not realized that Ptolemy was using calculations that underestimated the circumference of the earth by about 25 per cent. The map shows the world as contained within 180° of longitude, from the Canary Islands in the west through to Asia in the east. It creates a perception that Asia continues off the map to the right and thus 'round the other side' making a journey from the west appear viable. It seems highly likely that Christopher Columbus was aware of the 'new' Ptolemy map created in 1482. It may well have increased his confidence as he set out into the Atlantic on his momentous voyage just 10 years later. He was, after all, intending to sail to Asia not America.

Finally, it may be of interest to the reader that there is currently an American cartographer working with Ptolemy's original coordinates and following his advice to re-plot these. He is using the latest technology to recreate a twenty-first-century version of this magnificent map of the ancient world.

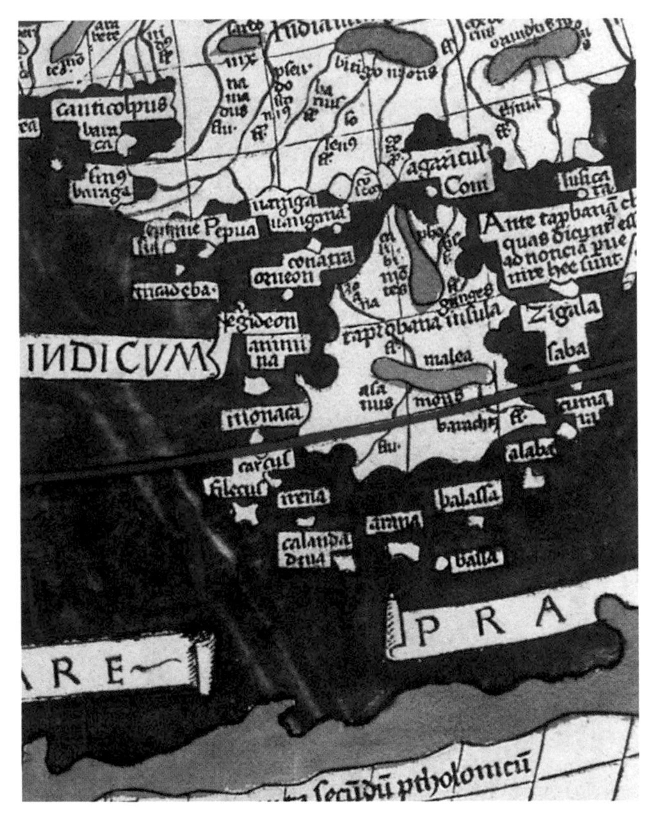

Terra Incognita

Ptolemy was meticulous and incredibly thorough but like all mapmakers he was limited by the information available to him at the time. One particularly noticeable feature is his presentation of the Indian Ocean as an enormous but enclosed mass of water – in effect an inland sea. The coast of China and that of east Africa are shown joining on to a southern continent stretching along and below the Tropic of Capricorn. Ptolemy doesn't speculate as to what might be there, simply recording 'terra incognita'. The myth of a southern continent would continue for another 1,500 years and although its 'location' would be gradually shifted south once the tip of Africa was rounded, the conviction that it existed remained firm. It was not finally dispelled until Captain Cook's voyage in the latter half of the eighteenth century, which established once and for all that Australia and New Zealand were islands and that there was no undiscovered landmass beyond.

Breaking the Frame

Although Nicolaus Germanus's re-creation of Ptolemy's map was true to the original text, he also thought it appropriate to make some small additions to 'update' it. One of the most significant of these was the inclusion of the North Atlantic and, as the extract at the bottom of p. 24 shows, this meant breaking out of the classic Ptolemic frame. The 'bump' that has been added does rather destroy the artistic symmetry of the whole map but the inclusion of the information on Scandinavia is significant. It seems to be the first time Iceland and Greenland had appeared on a world map. The former is shown fairly accurately, while Greenland is shown as a peninsula linked to Europe. Even with the additions, the information about this northern corner of Europe was far from complete, because Nicolaus is content to reproduce Ptolemy's orientation of Britain with Scotland at a 90° angle to England.

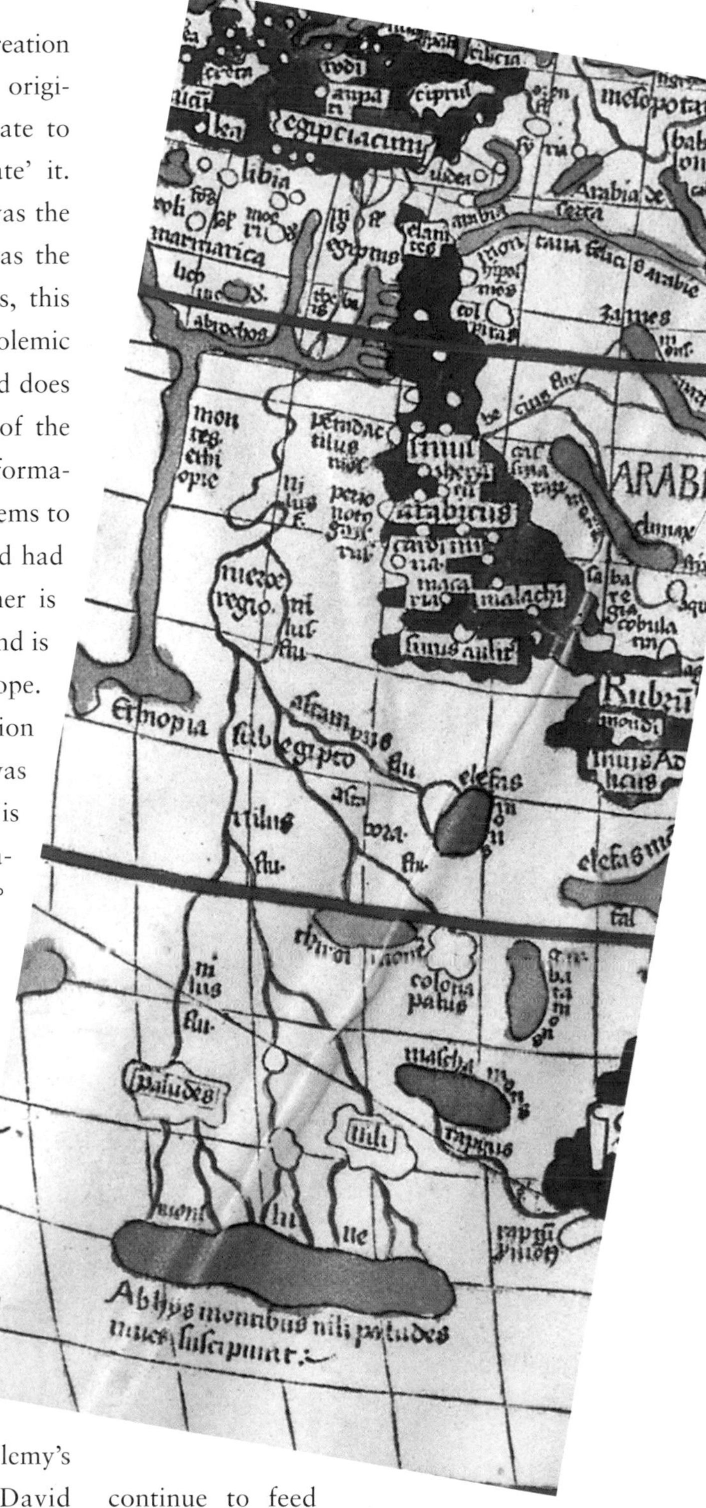

The Source of the Nile

While Ptolemy may have known little of northern Europe, his knowledge of the river that lay at the heart of his adopted homeland was extensive. The meandering path of the River Nile is shown in detail from its delta to its inland tributaries. It would be almost 1,700 years after Ptolemy's death that H. M. Stanley and David Livingstone finally established that the source of the Nile did indeed originate in a great inland lake close to the equator – as is represented here. The mountain ranges that continue to feed this and the surrounding lakes have different names today but none capture the magic of Ptolemy's *mont lune*, the Mountains of the Moon.

CASTORI ROMANORUM COSMOGRAPHI

tabula quae dicitur Peutingeriana. Recognovit Conrad Miller. 1887

SEGMENTUM II

II. 1. 2. 3.

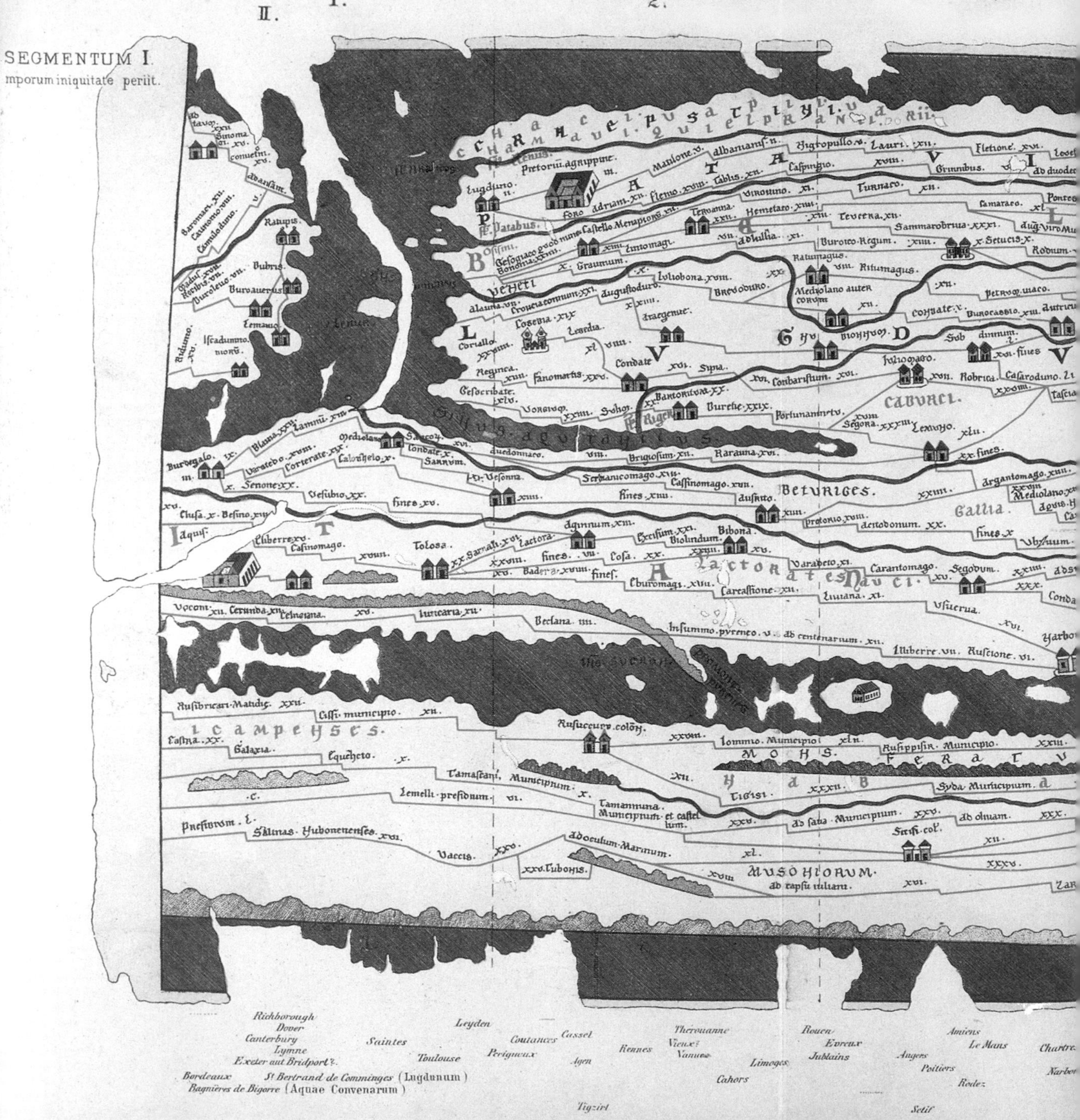

All Roads Lead To Rome

The Peutinger Table

By an unknown artist, *c.* 1265

All Roads Lead to Rome
The Peutinger Table

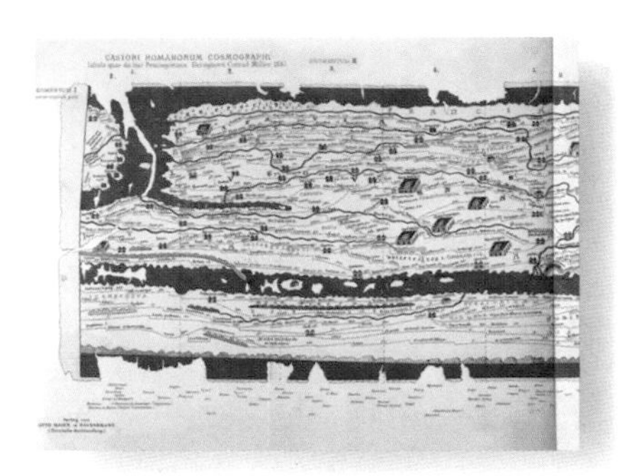

The phrase 'all roads lead to Rome' is used today to imply that whatever decision is taken the outcome or destination will be the same. However, at the height of the Roman Empire it was more than a figurative use of words, it was true in a literal sense. While there are many wonderful remains from the Roman period, no actual maps have survived and this one is as close as we can get to those used by the Romans.

This map's history is an example of the rôle 'chance' plays in what survives from the past. In 1508 a certain Konrad Celtes (or Bickel) died and left a map in his will to his friend, Konrad Peutinger, a citizen of Augsburg in Austria. Celtes had been the official librarian to the Emperor Maximilian and there remains some doubt as to whether he had acquired the bequeathed map lawfully. Peutinger himself was a leading scholar of his day and realized that the map was a medieval copy of an earlier Roman original. Despite the fact that this was his sole contribution, the map has become known, somewhat misleadingly, as the Peutinger Map (or Table).

It is an incredibly important document that has, and continues to provide, information about the Roman era. It is generally accepted that the 'Peutinger' copy was made around 1265 by a monk at a monastery in Colmar, a small town on the present France/Germany border. But what was he making his copy from? This is where the map's real significance lies, because it is believed that he was working from an original from the fourth century that has not survived. The unknown monk drew his copy on a long narrow scroll of parchment measuring over 22 feet (6.75 metres) in length but only 13 inches (34 centimetres) in depth. Whether these measurements corresponded to the Roman original can never be known but the assumption is that it was of similar dimensions and also came as a scroll for ease of storage and use.

In its entirety the scroll showed the Roman world from Britain to present-day Sri Lanka. The scroll itself was cut into 12 sections, only 11 of which have survived. The area shown here is on the second section. It is a very simple yet effective road map and has that functional, no nonsense style that informed so much of Roman design. In effect, it was a route map from all corners of the Roman Empire to Rome (which appeared at its centre) and it has been estimated that, in total, some 70,000 miles (112,000 kilometres) of roads are included.

The Appian Way, from Rome to the Adriatic port of Brindisi, formed the backbone of the Roman road network, and can still be walked today.

The practical approach suggests the map was intended to be used, because it addresses the two key questions posed by every traveller at any time in history: 'What's the next place I will come to?' and 'How far away is it?' Everything else has been made subservient to conveying this information in a clear and uncluttered form. The orientation is distorted and inconsistent and the scale varies. While the distances are usually given in Roman miles, local variations are included – for example, leagues are used in Gaul (France). Even allowing for the famous directness of the Roman road builders they were not as straight as presented here, but did this really matter to the traveller walking or riding along them anymore than it does to us today? Major rivers and mountain ranges are shown, if only in a crude form, but the traveller could be confident that if a road was shown there would be a bridge over, or a route through.

The whole map is elongated on an east to west axis. On this section France, North Africa and the Mediterranean are 'stretched' to allow the towns and distances to be written in. This also creates the impression that everyone and everything was eventually travelling to or from Rome – as in fact was usually the case.

In many ways it has a remarkable contemporary similarity to the route maps produced by modern motoring organizations. As here, places and distances are emphasized at the expense of other geographical features in the interests of clarity.

A Small Corner of England

When the map was created Britain would have featured on the first section, which has not survived. All that remains is this tantalizing corner of the south coast of England. From top to bottom the six 'twin towers' represent the present-day towns of Thetford (or Norwich), Richborough, Dover, Canterbury, Lympne and Exeter. A Channel crossing was then necessary to continue the journey to Rome. It is perhaps surprising that 'Camuloduno' (Colchester), which was one of the most important Roman towns in the area, is not accorded 'twin tower' status. Its name is simply written in alongside the road. This may provide a clue to this map's intended purpose. As with today's maps, the features that are chosen for inclusion and those that are omitted are determined by its function. For this reason the lack of significance given to a number of major legionary centres has led to the conclusion that this was not, primarily, a military map.

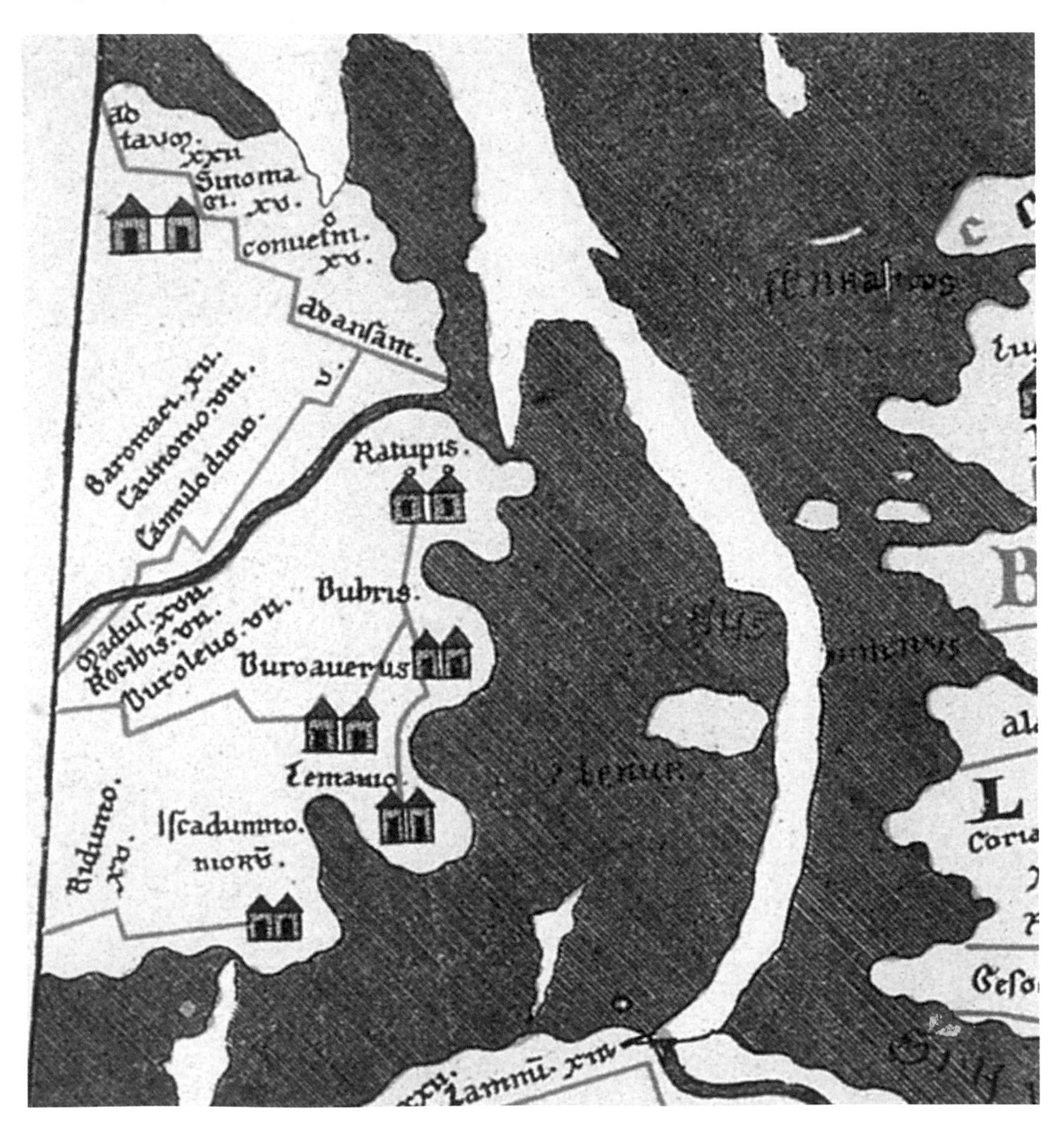

Roman Holiday

One of the most eye-catching features on this section of the map is the offset square structures, and their relative size clearly indicates their importance. With their gabled roofs and numerous entrances these were impressive buildings. The clue to their identity is to be found in the blue colour of the central enclosure, for these were spas. This is confirmed by the use of the Latin word *acquis* appearing before the name of the centre itself. The importance of bathing to the Romans is widely known but these are clearly more than the ordinary bathing houses and their significance on the map may well have been linked to claims made for their waters' medicinal or cosmetic powers. Aquis Calidis, the bottom left in this cluster, has been identified as the town of Vichy which continues as a spa town today. Its thermal and mineral waters continue to attract thousands of visitors every year. The range of mountains drawn to the south seem to be what we now know as the Masiff Central, an outcrop of volcanic rock through which the waters continue to take both their warmth and their minerals. There are other spas on the map but the cluster shown here almost conveys the sense of 'a centre' in the way clusters of ski resorts do today. Such an impression immediately returns us to the question of the original purpose and audience for this map. Could it possibly have been for wealthy Roman tourists planning their vacations?

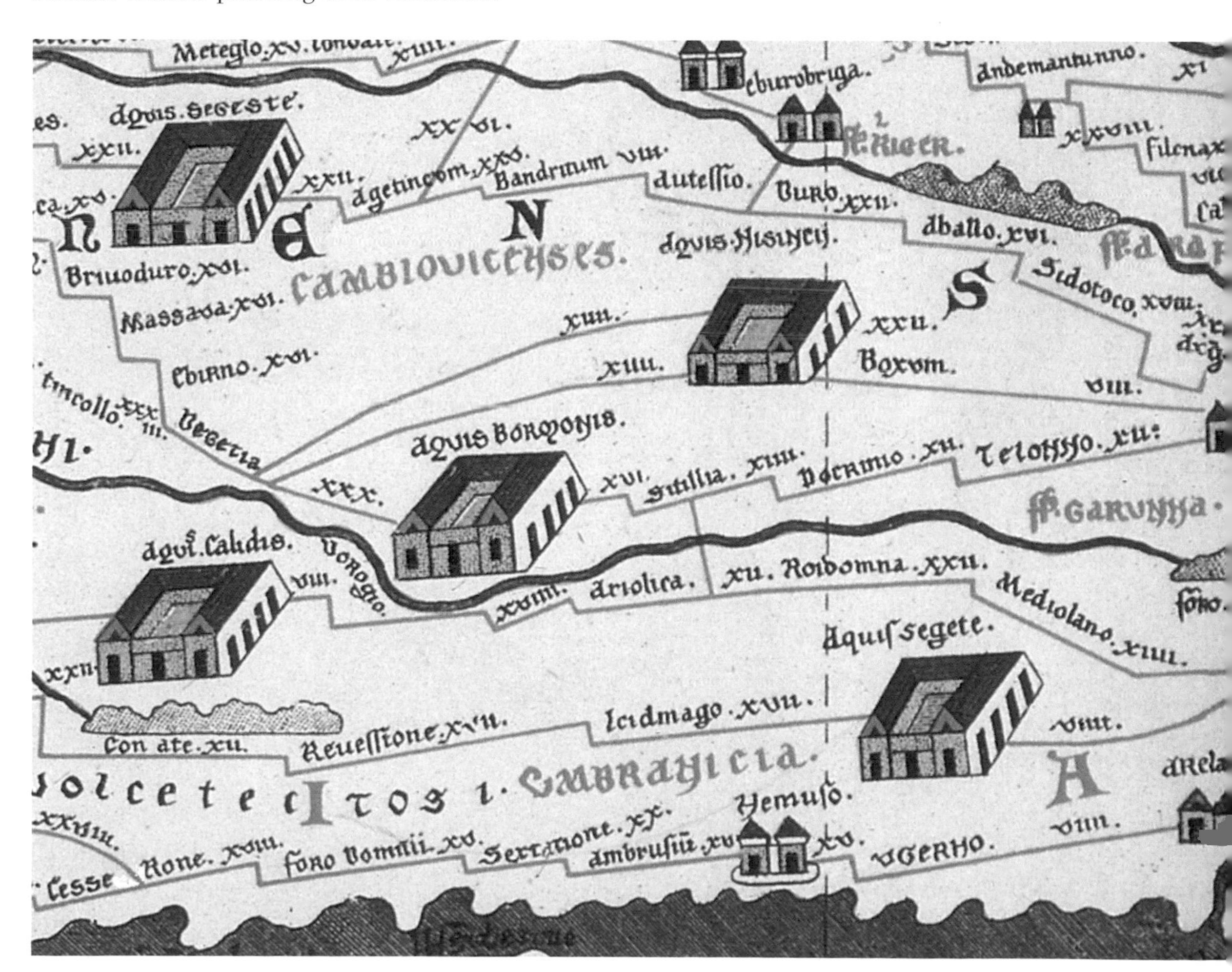

subsolanus
roma
grecia
egyptus

A Biblical Perspective
The Psalter
By an unknown artist,
1215–1250

A Biblical Perspective
The Psalter

A Psalter is a book containing psalms, and this tiny map, barely 6 inches (15 centimetres) high and 4 inches (10 centimetres) wide, was created in England to illustrate such a book some time between 1215 and 1250. It was never intended to be used for travel purposes and would have been of very limited value if anyone had tried to do so. Rather it is a symbolic map, which was designed to convey and reinforce certain messages, as, of course, were the words of the psalms that accompanied it. Even so, it does provide a wonderful insight into the way the geography of the world appeared when seen through the lens of Medieval Christianity. Despite its size it contains a wealth of detail, some geographically sound, some pure fantasy, and much in between.

The map is dominated by the figure of Christ omnipotently presiding over a world spread out before him, almost like a table. The stars of heaven provide the backdrop, while angels worship him at either side. The two dragons crouching in the dark at the bottom of the world represent another, darker kingdom.

The world itself is presented as a circle surrounded by sea and at its centre is Jerusalem. As with many maps the choice of what is placed in the central position is usually deliberate. Perhaps more significantly, and even subliminally, it also acts as the point from which other features are then viewed and related.

This map is orientated with East at the top. This ensures that the highly symbolic Garden of Eden appears in a prominent position just below the figure of Christ and with the sun directly in between. The somewhat pensive faces of Adam and Eve and the Tree of Temptation can be clearly seen in the Garden, which is enclosed by mountains. Five rivers flow out of Paradise and the familiar names of Ganges, Tigris and Euphrates can easily be read. While the latter two are reasonably accurate in location, the Ganges is clearly not.

The details on the map actually become more familiar to the modern eye if it is rotated clockwise by 90° so that North is at the top. It is then possible to recognize

Floor mosaic in the church at Tabgha by the Sea of Galilee, which was built on the traditional site of the miracle of the multiplication of the loaves and the fishes.

the fan of blue zigzags representing the Nile delta as it enters the Mediterranean Sea. The green of the Mediterranean can also be followed to the West where it flows into the sea that encompasses the whole world. North of the Mediterranean one can make out Greece and its islands in the Aegean Sea, and Italy, although France and Spain seem to have been rolled up together. In relation to its purpose this would have been of little importance to the map's creator. When we look to the south of the Nile, myth and legend rather than fact informs the features presented. The lack of knowledge of this region had led to the belief that the people who lived here were different in form. Those shown here, especially the ones with faces in their chests, would continue to feature on maps of Africa for several hundred years.

With its audience in mind the map gives over half of the world it represents to the Holy Land. It strives to make as many biblical references as possible and invites the viewer to make others. The Rivers of Jor and Dan can be seen flowing into the Sea of Galilee in which a large fish swims. Whether this is an indication of its role as a food source or an invitation to think about the story of Jesus and the loaves and fishes is not known. Perhaps it was both.

This map is almost certainly a copy of an earlier one but the identity of the person(s) who worked on it is lost forever. It is most likely that he undertook the work in a monastery or other religious house setting. Representations of the world through Christian teachings are collectively known as Mappaemundi of which only a small number survive. The one shown here is among the smallest of these, which makes the amount of detail contained so remarkable. The fate of the largest one, known as the Ebstorf Mappaemundi (some 11½ feet/3.5 metres across), is a reminder of just how precarious their existence has been over the centuries. It was destroyed in an air raid in 1943.

Moses Parts the Red Sea

The colour of the conical-shaped feature at the top right of the map has been specifically chosen for its literal association. This is indeed the Red Sea. As with several other features a 90° rotation of the main map makes it easier to recognize to our eyes. The Red Sea is a very significant location in the Old Testament. It is where Moses parted the waters to allow the Israelites to escape from the pursuing Egyptian army as they escaped to the Promised Land. Its prominence and colour on this map is clearly intended to invoke this story from the Book of Exodus. A closer look at this section of the map reveals that the Red Sea is actually 'split' with a dry passage shown between. Could the dark blue line on the larger section be intended to represent the wall of water held back by Moses' command until the Israelites had crossed and then released on the Egyptians as they attempted to follow?

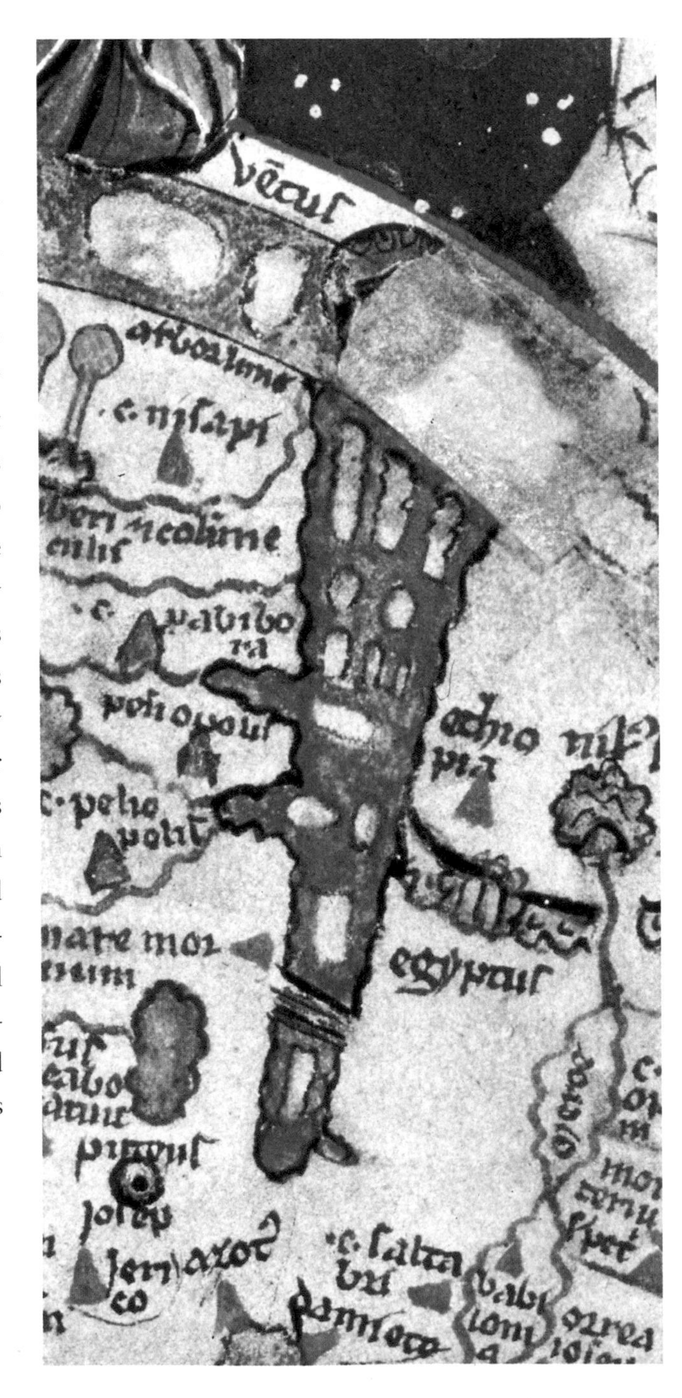

Gog and Magog

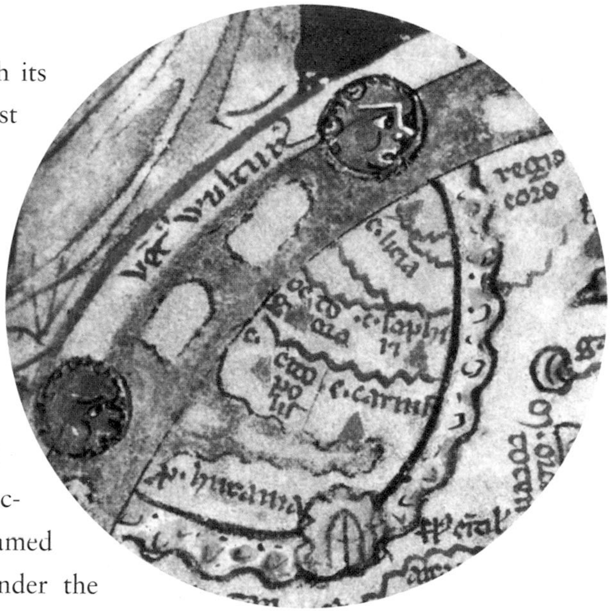

This mountain-like walled enclave with its single gate represents one of the most apocalyptic stories from the Old Testament. The prophet Ezekiel had warned of the day when Gog, the chief Prince of the Land of Magog would sweep down from the North and devastate the land foretelling that the Day of Judgement was near. The gates shown were known as the Gates of Alexander and were reputed to be all that held these destructive forces back. The gates were so named because legend had it that it was Alexander the Great who had built them in a narrow pass to seal in the fierce tribes from the north. But it was also part of the legend that these could not last forever and Gog and the Magog would one day break through. It was a story that all Christians of this period would have been familiar with and this section of the map is intended to convey this threatening and ominous presence.

Noah's Ark

The story of a gigantic flood and a boat in which a small group of chosen people and animals survive appears in many cultures around the world. Not surprisingly it is the biblical version that is portrayed on the Psalter map. The search for the ark itself has fascinated generations and continues today. This map has played its part in informing the search because, as can be clearly seen, the ark is shown marooned high between two mountain peaks with the words 'arca noa' and 'armenia'. Just above 'arca noa' can also be read the word 'Herat'. This is accepted as being Mount Ararat located on the present-day border of Turkey and Armenia. In the final years of the twentieth century satellites were used to take detailed photographs of the whole area and the images released caused great excitement in some quarters. However, in the thirteenth century when this Psalter map was created, there was no controversy or doubt about the ark's existence or location and it would have been drawn on with the same conviction as any of the other features.

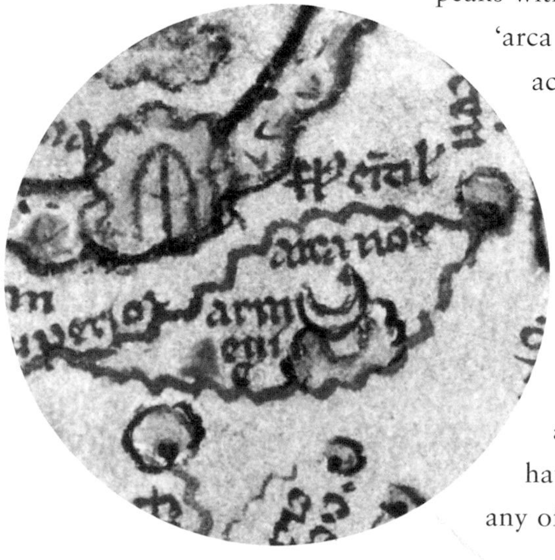

SCOCIA: ULTRA MARINA
hec et albania dicta est
WALLIA
london
DORSET
CANCIA
Chanet
AUSTER
ORIENS

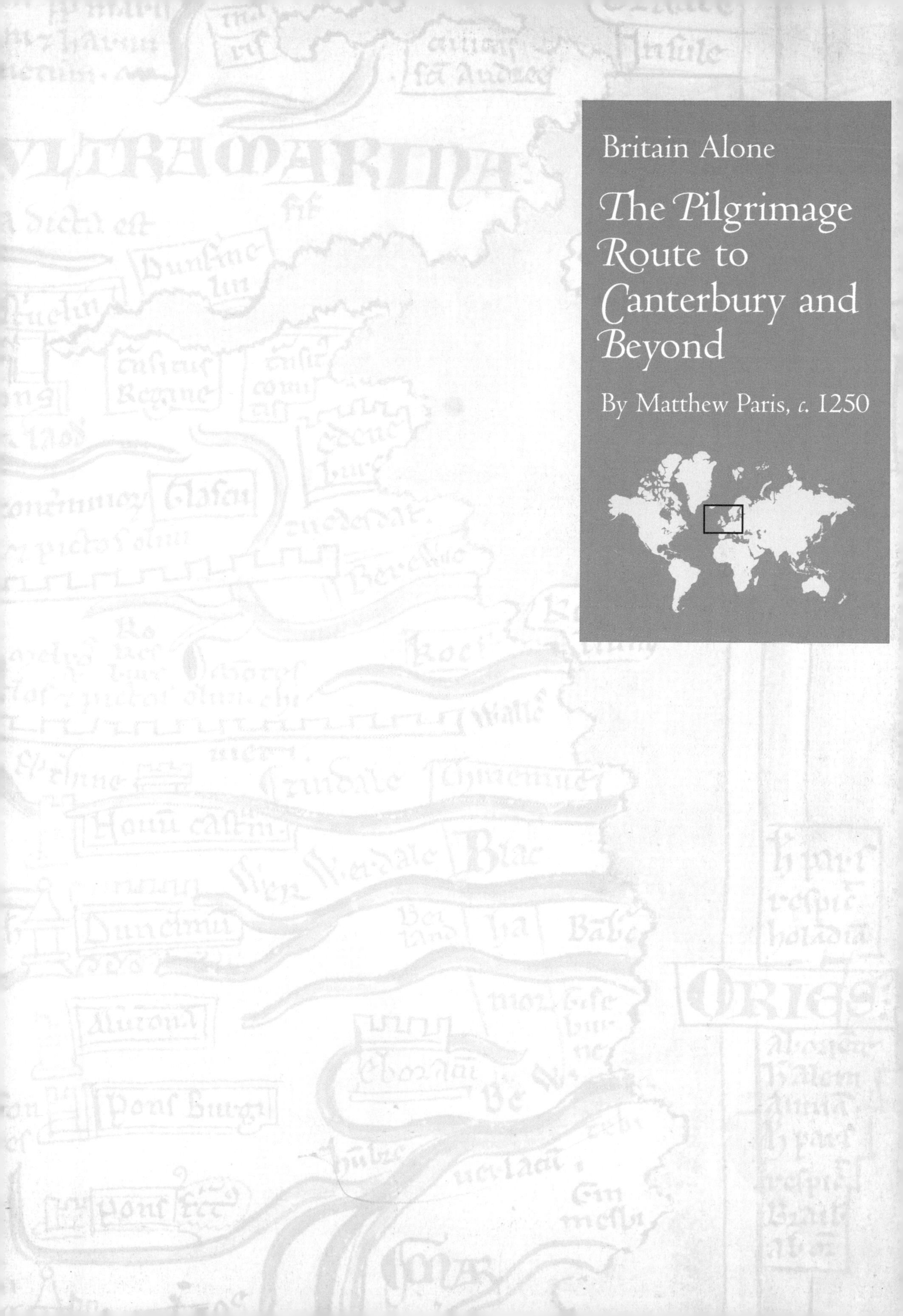

Britain Alone

The Pilgrimage Route to Canterbury and Beyond

By Matthew Paris, *c.* 1250

Britain Alone

The Pilgrimage Route to Canterbury and Beyond

At a first glance this map seems familiar yet puzzling to the eye. On further examination it becomes apparent that it is indeed a map of mainland Britain but one that is somewhat distorted. This is due to its intended purpose rather than any lack of information or skill on the part of its creator.

The map is one of the earliest surviving detailed maps of Britain and was drawn by Matthew Paris around 1250. Matthew is now regarded as the major historian of his time and his works continue to form the basis of our understanding of this turbulent period of British history. He was born in about 1200 and died in 1259, and during his life Magna Carta was signed, Prince Llewellyn drove the English out of Wales, Henry III became king at just 9 years of age, crusades were undertaken, Cambridge University was formed – and more.

Matthew had become a monk at St Alban's abbey in 1217. This was an ideal placement because St Albans took its role as safekeeper and recorder of knowledge seriously and he was able to work alongside more experienced scholars in his early years there. As he gained seniority he expanded his role and wrote annual chronicles from 1239 to his death, in which he recorded and commented on major events. Crusades and uprisings appear alongside storms and the price of grain but he was not content to simply record. He also felt it was his role to comment critically on the events he wrote about and neither pontiffs nor kings were safe from his pen. The fate of Thomas à Becket, who as Archbishop of Canterbury had been murdered following his opposition to the King's attempts to control the clergy in the previous century, had clearly not acted as a deterrent! Neither did it act as an obstacle to royal approval for he was on friendly terms with Henry and he also undertook an 18-month-long mission to the King of Norway on the Pope's behalf to enlist support for a planned crusade. Alongside his abilities with the written word he was a talented artist and illustrated many of his works himself.

A close look at the map reveals there is a very definite 'thread' that runs vertically down the centre with towns and religious houses drawn diagrammatically and named from Newcastle in the north to Dover in the south. To maintain this vertical perception Dover has been placed directly below London rather than to the east of it and, as a result, the Thames has been diverted southwards so that it flows into the Channel. The main aim of this document was not geographical accuracy but to act as a route map for the many pilgrims making their way from the different parts of Britain to either Canterbury or Dover before continuing their pilgrimage to the Holy Land. One might question the absence of roads if this was its intention but roads were not to appear on maps of Britain for at least another 300 years. Much more significant, and useful, was to know what the next town or religious house was along your route. By 1200 the remnants of the Roman roads continued to provide the basic network,

John Lydgate and the Canterbury pilgrims leaving Canterbury, from The Siege of Thebes, *in which Lydgate presents himself as one of Chaucer's pilgrims and is asked to tell a tale.*

supplemented by well-trodden track ways which often changed course according to the season. For the thirteenth-century pilgrim a town or abbey meant food and shelter, relative safety and the knowledge that there would be a bridge or ferry across any river. A traveller today hardly notices the numerous rivers and streams crossed on a journey but in Matthew's time these formed a major hindrance to progress and this is reflected in the prominence given to them on this map.The religious houses can be distinguished from the towns by their stylized spires, and it was under one of these that this map was drawn. The spire to the north of London is the Benedictine abbey of St Albans – *cenobium Sancti Albani*, Matthew's home.

A Little Help From the Romans?

The two walls the Romans built across mainland Britain are clearly, even prominently drawn on this map. As a leading scholar, Matthew would certainly have been aware of the Antonine Wall in the north and the more substantial Hadrian's Wall a little further south. However, he drew his map around 1250, over 1000 years after these were completed and some 850 years after the Romans had left Britain. Considering the purpose of his map the walls seem somewhat superfluous. While his reasons for including them will never be known their presence does lend support to the idea that he may have been working from an earlier Roman map. If the position of the walls was copied from this it is interesting to muse over what else he chose to copy and what he chose to leave out as unnecessary for his own requirements.

Matthew's, and indeed the map's original home at St Alban's abbey, also has significant Roman connections, because St Alban was put to death by the Romans for his Christian faith and thus became the first British martyr.

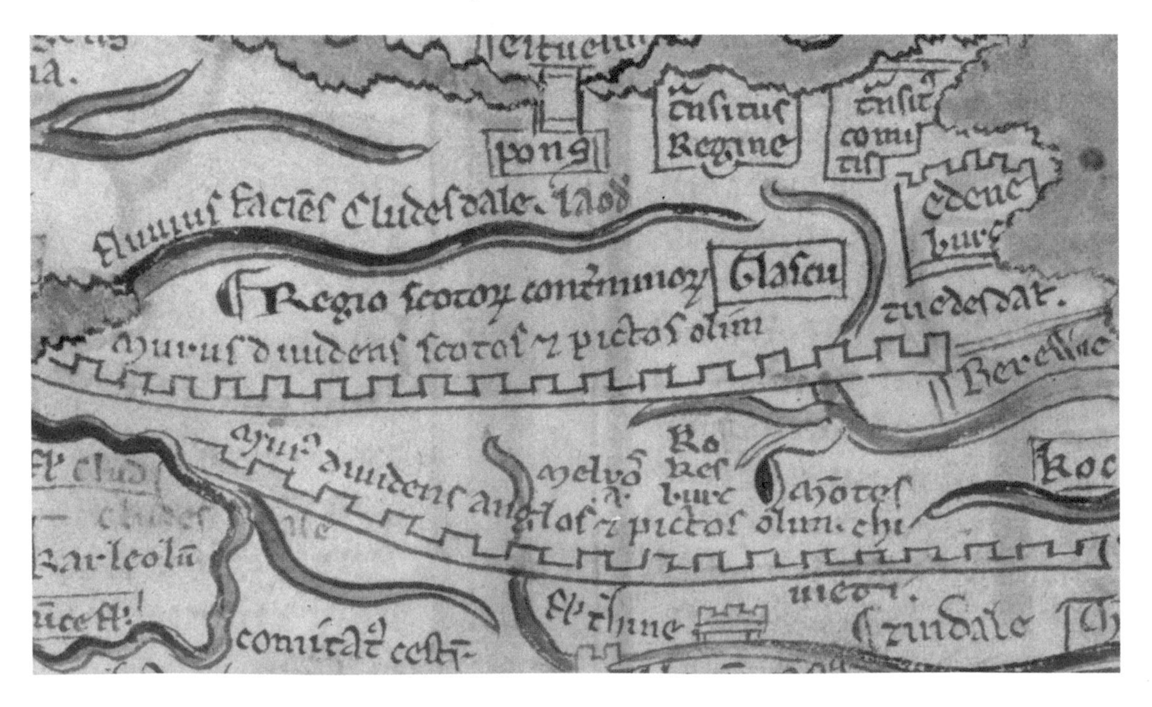

Reliable Distortions

The area between London and Dover is distorted as well as being rather small for the amount of detail Matthew sought to include. The two rivers marked are the Thames and the Medway and both have been rotated some 90° south from their true orientation. As Matthew indicates, a pilgrim leaving London would have headed for Rosa (Rochester), the lowest crossing point on the River Medway before reaching Cantuar (Canterbury). Chaucer's amusing yet informative tales of the pilgrims heading for this pilgrimage site are well known and like his characters many would have ended their journey here. For others it was an important stopping place but they would have then continued on to Dovia (Dover). The fortified image reflects the immense castle that had been built there to protect what was then, as now, the main port of entry into Britain.

Matthew would have known the London he drew at the heart of his map very well. He was often a guest of the royal court and would have visited the new and grand Westminster Abbey that Henry was building. Whether he actually saw the whale his Chronicle records as swimming up the Thames is less certain!

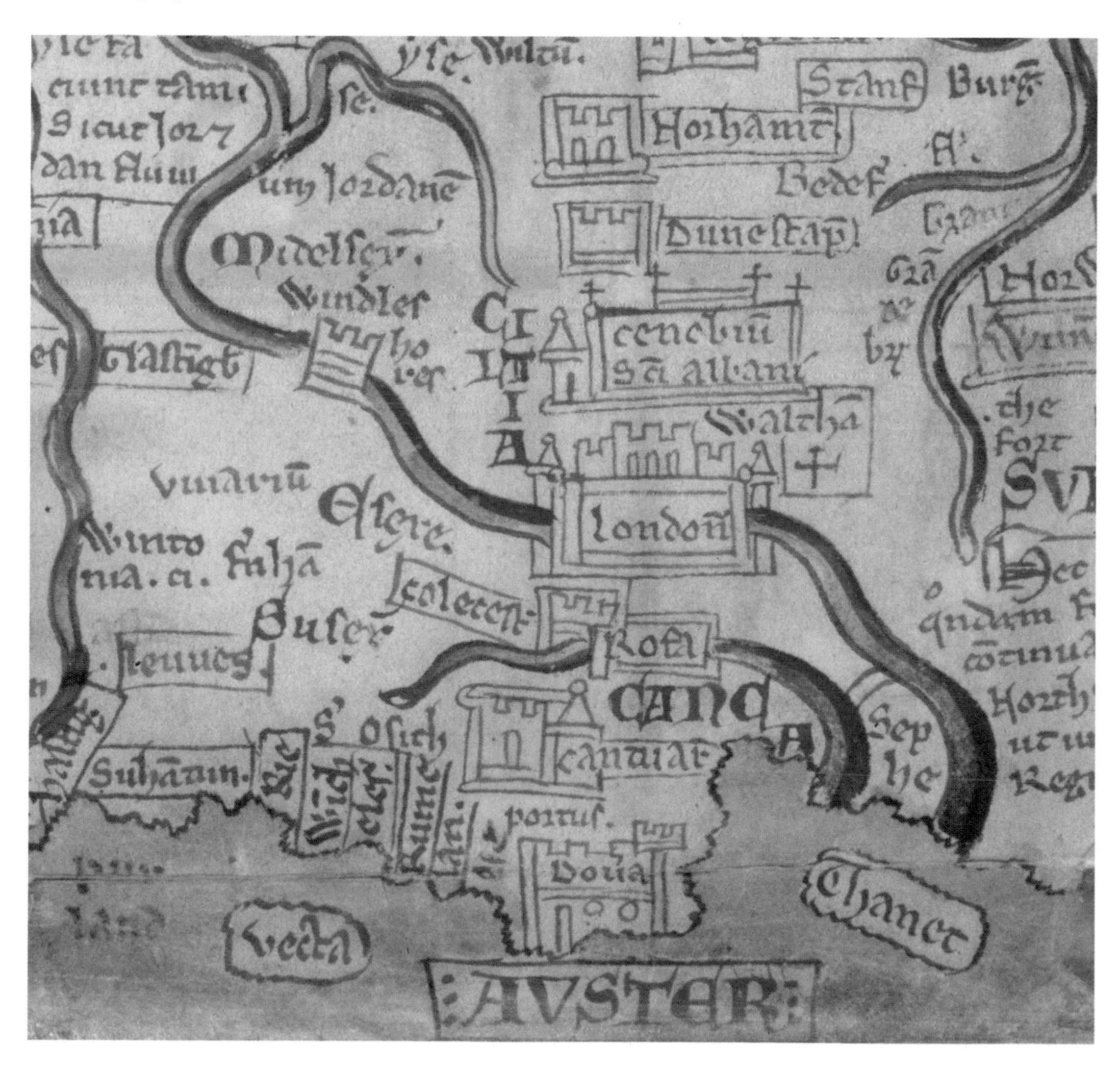

...imus contra omnes gentes warantizabimus. In cuius rei testimonium hu[ic]
nre sigill nrm apposuimus hiis testibus Ricardo ffetz John Robto Rylvtymbre
Selley et aliis. Dat apud Certesey in vigilia Sci Mathei apti Ao rr Henr Sexti post

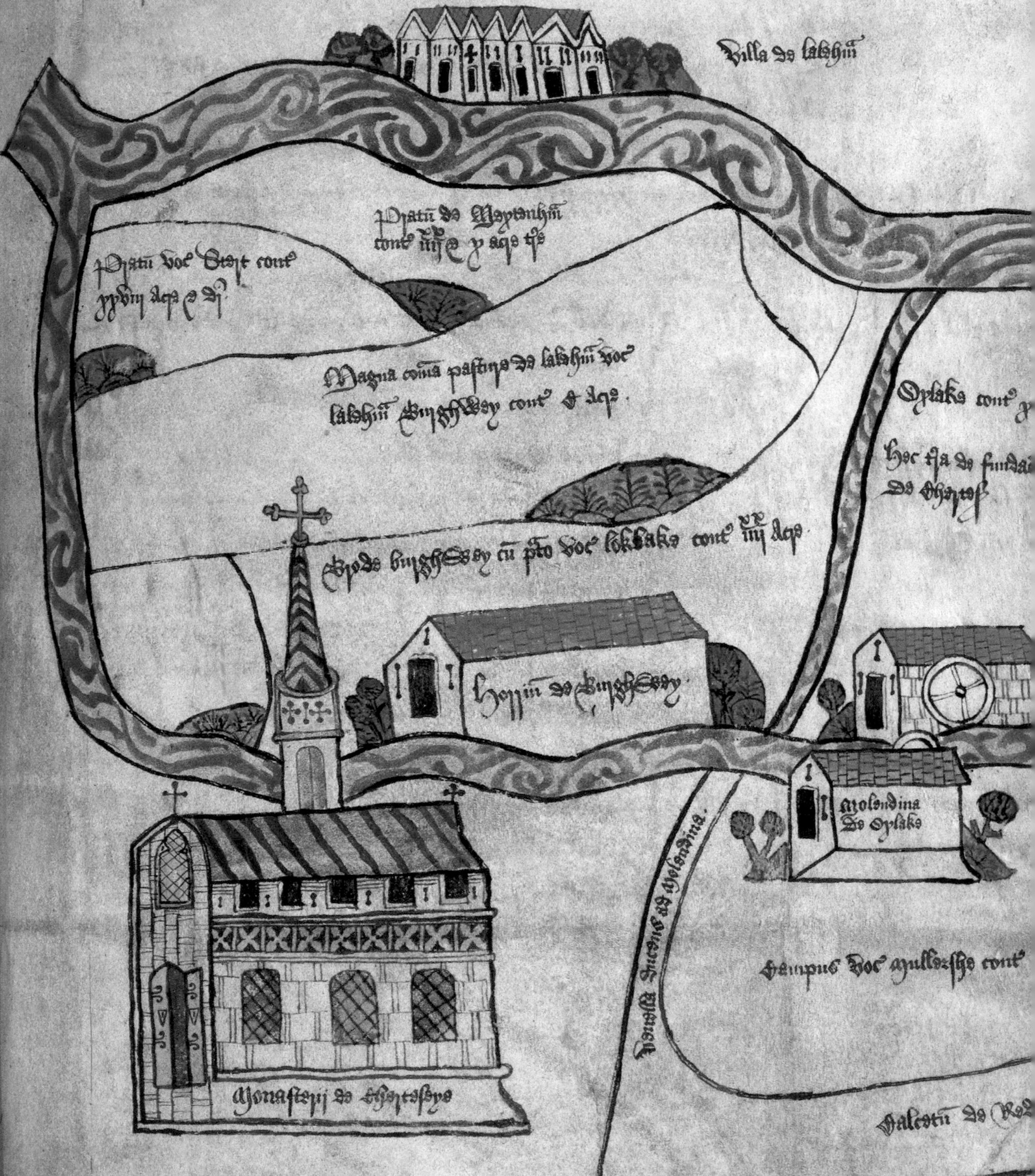

Villa de Laleham

Pratu de Maytonham cont iiij x acs ts

Pratu voc Stept cont xxviij acs et di

Magna comuna pastura de Laleham voc Laleham Burghwey cont [illegible] acs

Oxlake cont

Hoc t[er]ra de fundac de Chertesey

Prode Burghwey cu pto voc Lokslake cont iiij xx acs

Hospiciu de Burghwey

Molendina de Oxlake

Venella ducens ad Molendina

Campus voc Mullesrsh cont

Monasterij de Chertesey

Walcotu de R...

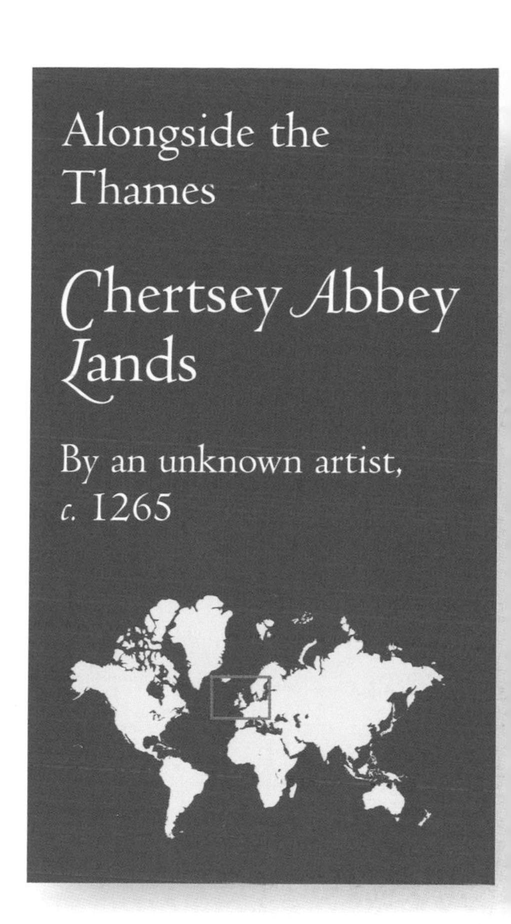

Alongside the Thames

Chertsey Abbey Lands

By an unknown artist, *c.* 1265

Alongside the Thames

Chertsey Abbey Lands

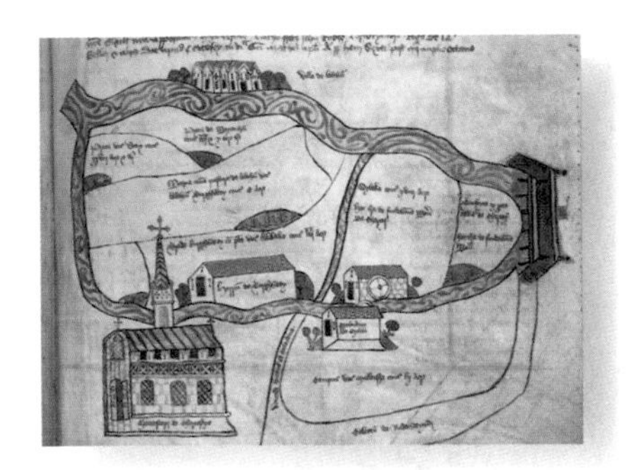

This map was drawn to resolve an argument between the Abbey of Chertsey and the local farmers about who had the right to graze their animals where. By any definition this was not a major historical event! But most people's lives consist of many small events whose significance is very real to them, and so it would have been to the farmers involved in this dispute. To assist the abbey's case a senior monk, possibly the Abbott himself, asked for this map to be drawn up to clarify the situation. We are fortunate that it survived for, although simply drawn in the manner of a sketch map, it gives us as good a sense of contemporary life as any of the more famous ones of the period. Although particular to this part of Surrey, the scene presented here would have been found with only small variations right across England.

The massive, stone-built church of cathedral dimensions would have dominated the rural landscape. As the map portrays, this landscape consisted of the cluster of cottages that formed the village of Laleham, the water mill, its millrace, the tithe barn, the wooded copses and the common land. Around these were the meadows and fields crisscrossed with narrow tracks rather than roads. While it may sound idyllic, life was short and hard for the people who worked on the abbey's land and they paid in kind for the right to do so.

As the map shows, Chertsey Abbey was situated on the banks of the River Thames. This was the major thoroughfare into London in the medieval period and barges carrying goods and people would often have stopped here as they made their way upstream to Windsor. Interestingly, the abbey's most famous visitor was dead when he arrived. Being King of England was not a secure position during the Wars of the Roses and Henry VI was murdered in the Tower of London in 1471, probably on the orders of his successor Richard III. His body was carried by boat to Chertsey Abbey and buried in the church featured on this map. This event took place only a few decades after this map was drawn.

Originally, the riverside location would have been one of the key factors that influenced the newly converted Saxon Chief, Frithwold, to found a Christian community here in AD 66. Over the succeeding centuries, however, this setting made the abbey vulnerable. In the era of the Vikings one raid saw the Abbott and 90 monks murdered. Its royal links were firmly established when King Edgar rebuilt the abbey in the tenth century. Over the next 500 years it was extended, redesigned and rebuilt until, like monasteries across the country, its fate was determined by Henry VIII's failure to produce a male heir. The 'dissolution of the monasteries' was one of many consequences that resulted from Henry's decision to divorce Catherine of Aragon and his subsequent expulsion from the Catholic Church.

Pieces of stone and other artefacts from the abbey survive in local buildings and have also been recovered during excavation work.

The impact on Chertsey would have been dramatic. The abbey was demolished and its highly prized stone, timber and stained glass transferred to Weybridge where it was incorporated in a new palace Henry was having built. Some lesser stone would certainly have been used in local buildings, where it probably remains today. However, the monks' 'curfew' bell did remain in the area and now rings in the local church of St Peter's. The dissolution inventory of 1537 also survived and shows a total value of £659 15s 8d – an enormous sum at the time. It went straight into the royal coffers. The local people had a new landlord but the pattern of their lives would have continued much as before.

The Benefits of a Bridge

Rivers once created very significant barriers to travel and the bridge at Chertsey was an important one. The sketch shows it was wooden, and it would have been built and owned by the abbey, which could charge a toll for using it. A common practice was to make a deal with a local man to maintain the bridge in full or part payment of his rent or dues on his land, and this was most likely the situation here. It is interesting to note the road leading up to the bridge from the south is termed a 'causeway'. This usually indicates a raised path over waterlogged land and would have been necessary to cross the Thames floodplain either side of the bridge. It is possible that the name 'Redewynd' may have been the villager responsible for its upkeep when this map was drawn.

The benefits to the abbey of its bridge were, however, much greater than the total of the tolls it might collect. In the Medieval period, even more so than today, the location of a bridge affected communication patterns over a wide area, 'funnelling' roads and paths, and thus trade and travellers towards it. Chertsey Abbey, like most religious houses, offered overnight hospitality and accommodation to travellers – and charged accordingly.

A Double Mill

The abbey was very wealthy and owned land in Henley, Egham, Epsom, Chobham, Weybridge and even as far away as Tooting. The rents and dues from those working this land were a major source of the abbey's wealth and the senior monks were astute businessmen.

One of the most profitable sources of income came from the water mill shown here. Named as Oxlake Mill, closer examination reveals it is, in fact, two mills standing either side of the stream. The top of the water wheel belonging to the nearest building is just discernible above the roof. A double mill is indicative of the amount of milling of wheat, barley, rye and possibly other crops that was undertaken. One of the conditions attached to the rental agreements with tenants would almost certainly have been a requirement to have their milling done here. Besides charging the tenant for this, the abbey would also take its 'tithe' or tenth share, which was probably stored in the large Burghwey Barn shown on the map. And then there was the miller himself! As Chaucer's *Canterbury Tales* (written at the end of the previous century) clearly reveal, millers were notorious for their devious practices when weighing sacks and pricing their services – as well as for their bawdy stories.

One might imagine the conversations that took place along the track way shown here as the unfortunate tenant farmers left the mill with their remaining share. A 'good moan' may have made them feel a little better but there was nothing they could do about the abbey's monopoly and they would be back again after the next harvest.

Chapter Two

The Quest for Riches

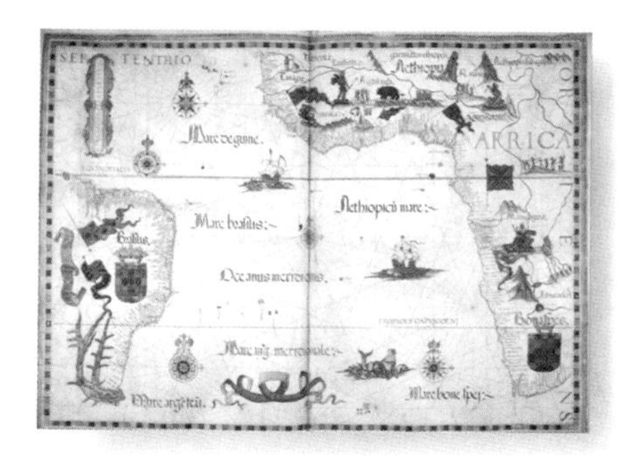
AFRICA

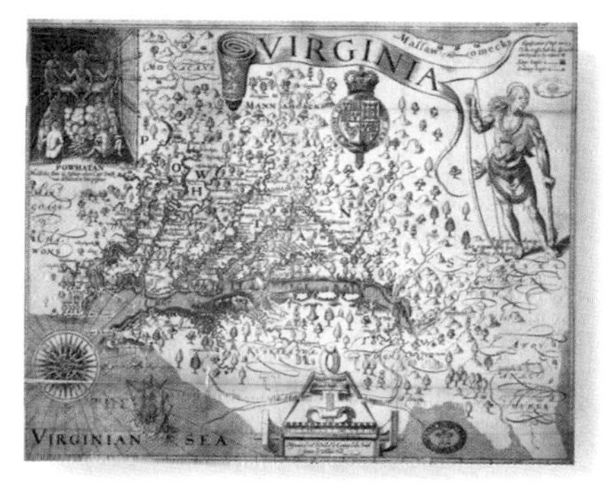
VIRGINIA
VIRGINIAN
SEA

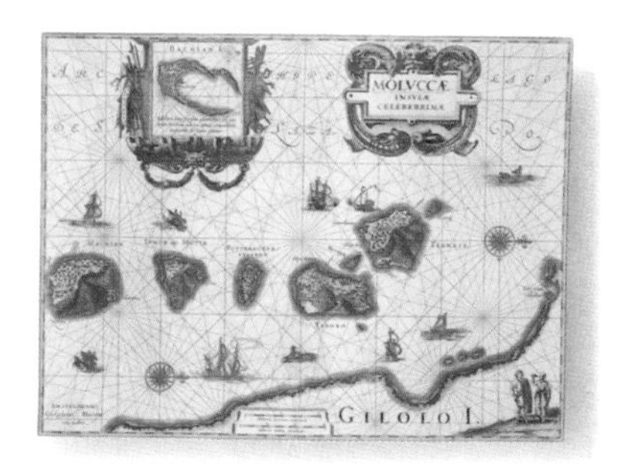
GILOLO I.

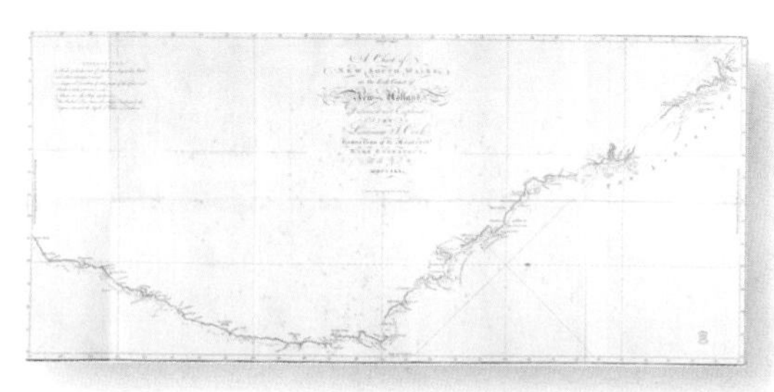

For every map there has to be a starting point. There needs to be some information to include and someone has to collect it. On some occasions, as in the case of John Smith (pp. 66–71) or James Cook (pp. 78–83), these two roles were performed by one and the same person. At other times the mapmaker was dependent on information being brought to him. His skill lay in checking its validity and incorporating it into the existing knowledge to present an updated or up-to-date version. Willem Blaeu's map of the Spice Islands (pp. 72–7) and Diego Homen's map of the West African coast (pp.60–65) are good examples, as there is no record of either having visited the locations their maps portray.

Sometimes, as was illustrated in the previous chapter, we don't know who the actual mapmaker was but this does not negate the contribution he made. The exquisite map of Tenochtitlán that opens this section (pp. 54–9) was almost certainly copied from an original drawn by an Aztec, or Mexica, cartographer. The perspective suggests that he sat working high on the hill overlooking his city. He would not have known that he was creating a record of a place that was in the final years of its existence. Without his work our understanding of this civilization would have been so much poorer.

The relationship between the mapmaker and the information he has to work with is a very interesting one. As the European powers vied with each other to explore the world known to them in the fifteenth, sixteenth and seventeenth centuries this relationship was particularly important. It was the period during which vast areas of the world were represented in map form for the very first time. The use of the word 'discovered' is problematic when used to describe the voyages made by Europeans to the Americas, Australasia, Africa south of the equator and to parts of Asia. The people who were living there certainly knew they, and their part of the world, existed! What did happen in this period was that all the continents of the world were shown in relation to each other for the first time.

But what motivated individuals to leave the safety of their known world? They were literally going 'off the map', something that is almost impossible for us to contemplate today with satellite technology enabling a position to be plotted in even the most remote locations. Was it simply a desire to become rich?

The 'quest for riches' has always been the driving force for some people and when the adventurer Hernán Cortés recruited his conquistadors to accompany him to Mexico it would certainly have featured. When the Portuguese explorer Bartholomeu Dias sailed south along the West African coast he too would have been seeking to fill his ship's hold with goods that would make a healthy profit when he returned to Lisbon. And then there was the colonist John Smith. The promise of becoming wealthy would surely have formed at least part of his agenda when he joined the Virginia Company and set sail for America.

But does the promise of wealth provide a sufficient argument to explain the actions of these men? Was it not also something to do with a certain type of personality that has always existed in society? No one knows why some people feel the need to undertake challenging and dangerous ventures. In the early years of the twenty-first century these have included walking alone to the North Pole and attempting to fly around the world in a hot air balloon. When interviewed they often seem to find it difficult to articulate and invariably refer to some form of 'inner drive' that needs to be satisfied. Without similar individuals in the past the information that enabled the mapmakers to 'complete' the map of the world would not have become available when it did. But similarly, many of the notorious events that also took place would not have occurred.

The Spice Trade in the Moluccas, by Guillaume le Testu, 1555. Control of the Spice Islands was fiercely contested by the various European powers for the riches that could be made in the spices that grew there.

Certainly, as one of the examples included here clearly shows, not all of those who sailed into the unknown were motivated by wealth. The Royal Navy Captain, James Cook, did no more than open his sealed orders and follow the instructions therein. The bravado found in Cortés' and Smith's accounts of their expeditions are absent from his logbook, which is generally modest and unassuming.

Two further points need to be remembered. First, while we use the names of the leaders to refer to their expeditions none undertook them alone. The conquistadors who accompanied Cortés, the would-be colonists who travelled with Smith, and the crews of all the various ships, made their own, often unacknowledged, contribution to the tales that unfold in the following pages. Nothing could have been achieved single-handedly. Many died in their efforts, and if it had been simply the desire for riches that spurred them on few would have died fulfilled. However, if to some small degree, they shared the desire to satisfy an 'inner need' their many experiences may well have achieved this end.

Secondly, the information brought back by these voyagers was assembled into a European view of the world. From their position it could not have been otherwise. As we look at the maps several centuries later an awareness of this perspective enables us to realize it could have been different from what seems so familiar to our eye today.

As the following pages show venturing 'off the map' was an exciting and very hazardous business. For once you had done so no one knew where you were.

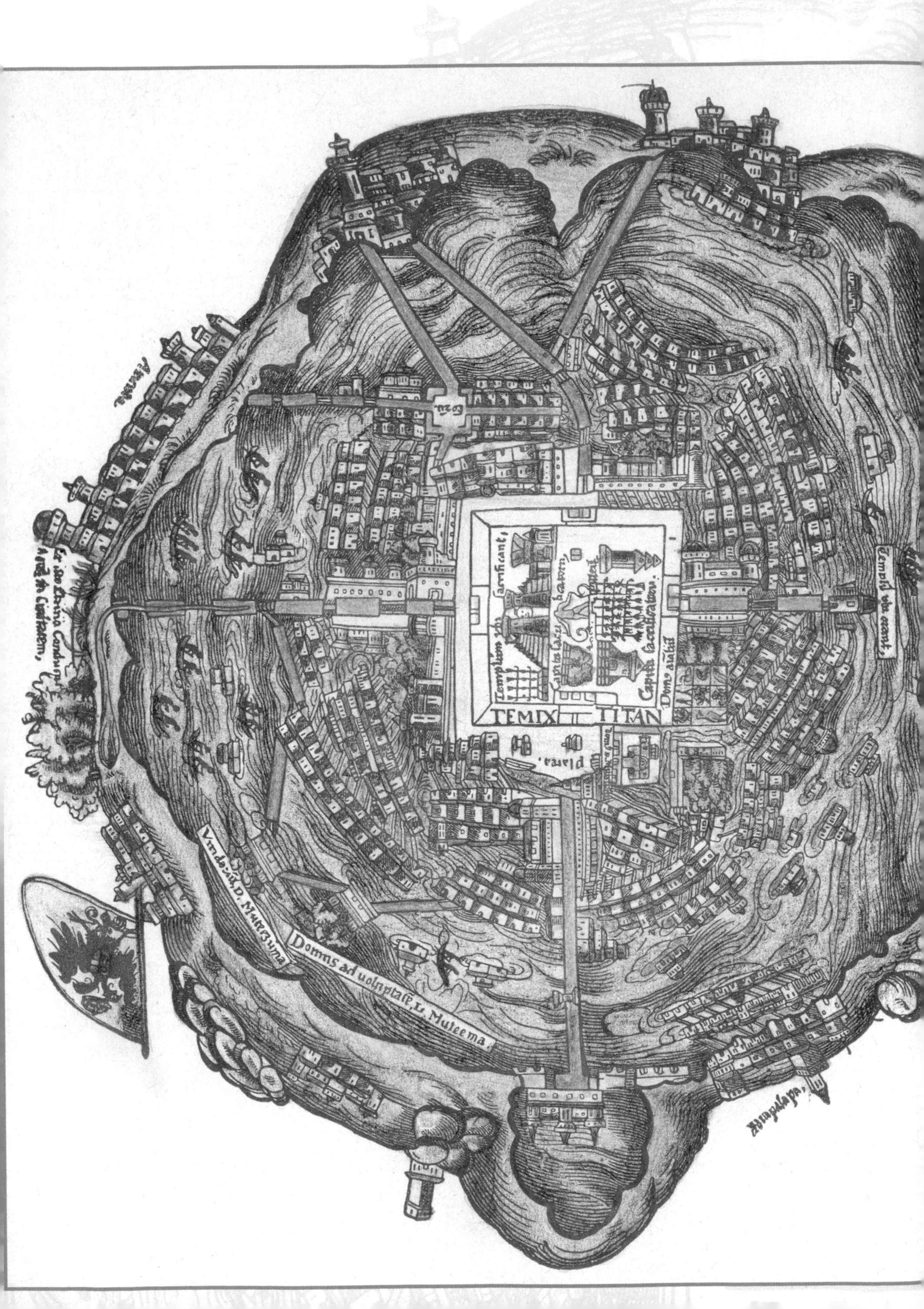
TEMIX TITAN

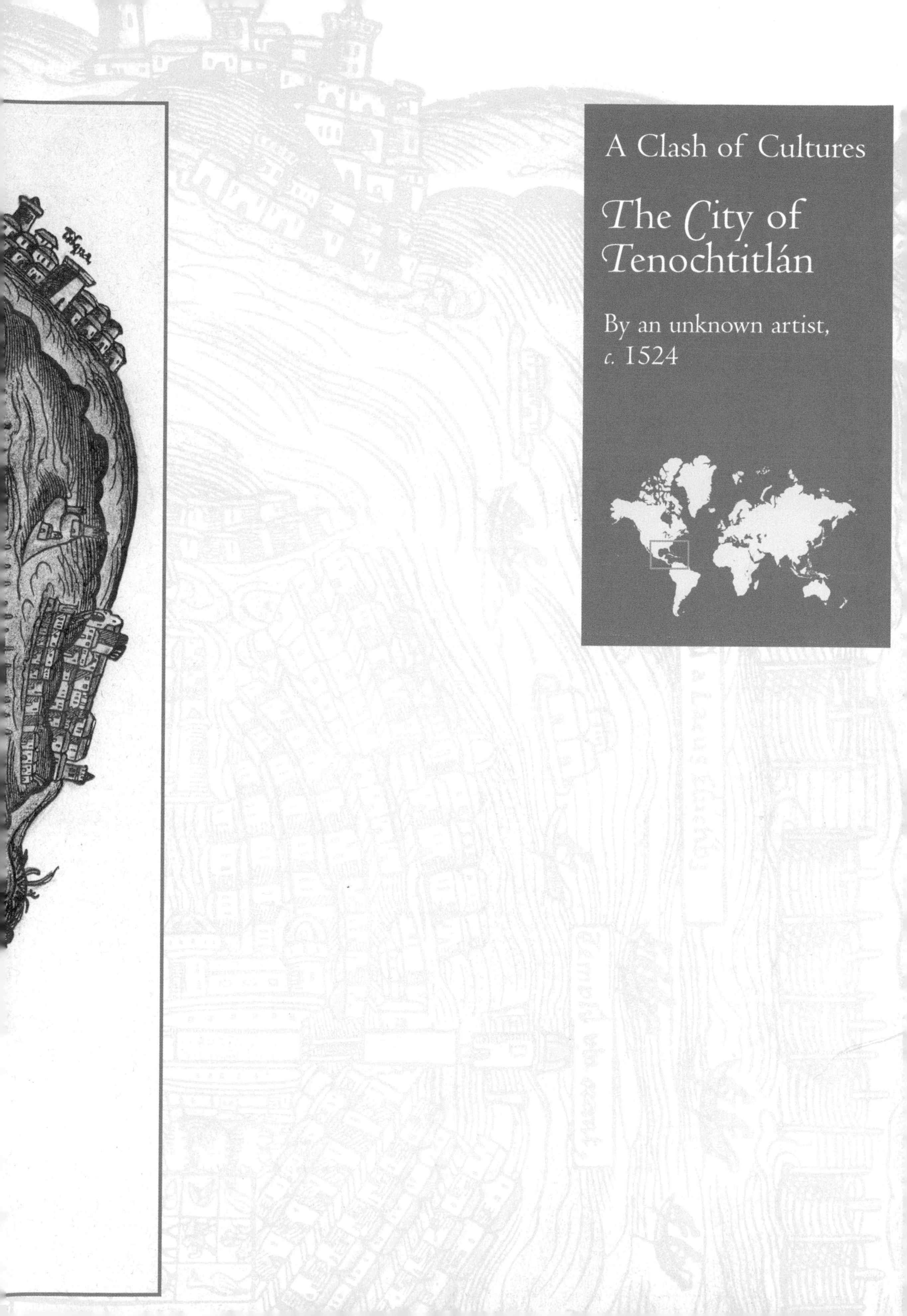

A Clash of Cultures

The City of Tenochtitlán

By an unknown artist, *c.* 1524

A Clash of Cultures

The City of Tenochtitlán

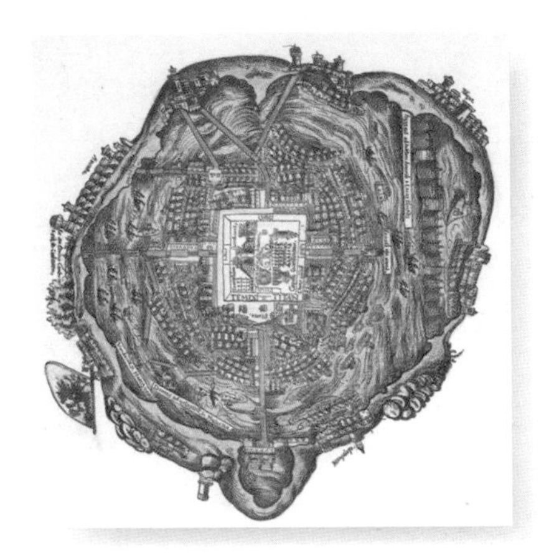

This beautiful and detailed map belies the barbarity and destruction that took place in the setting it portrays. For this is a map of Tenochtitlán, the capital of the Aztec or, more accurately, the Mexica Kingdom. The map was sent to the King of Spain by Hernán Cortés in 1524 but is almost certainly copied from an original indigenous source, the Latin text clearly being added for the benefit of the King. It captures the fascinating layout of the city and shows that Tenochtitlán was built on an island in a lake with causeways connecting it to the mainland. Each section of houses was separated from its neighbour by canals and linked with many small bridges. It is not surprising that when the first Europeans saw it they made comparisons with Venice. At its heart lay a large square – the Sacred Precinct. Recent archaeology has revealed this to have been some 4,360 square yards (4,000 square metres) in size surrounded by a high stone wall known as the Serpent Wall (the coatepantli) due to its decoration. This was where the Palace of King Montezuma II and the great pyramid Teocali (temple) stood and where ritual sacrifices were made to Tlaloc, the god of water and Huitzilopochtli, the god of the sun and of war. The square also housed the famous ball court. It was believed to be the centre of the world.

By the time this version of the map was drawn the city it portrays was no longer in existence. It had been looted and destroyed; its ruler murdered and thousands of its citizens killed in a bloody battle with Cortés's men and their allies.

Cortés was one of the numerous European adventurers who were drawn across the Atlantic by tales of fame and fortune. He had already been involved in the Spanish conquest of Cuba but his ambition was unfulfilled and in 1519 he led a force of some 450 men – the conquistadors – into mainland Mexico.

In a little over six months he was in a position to look down on the city on the lake shown on this map. There was an Aztec/Mexica legend that told how their white God-King Quetzalcoatl would one day return from the East. Montezuma and his priests seemed uncertain as to whether Cortés was the fulfilment of this prophecy and this ambiguity proved catastrophic. They welcomed the white men on their strange beasts (horses were not known in this part of the world at the time) into the Sacred Precinct. However, Cortés feared a trap, took Montezuma and his entourage captive and attempted to rule the city through him. Within months the people of Tenochtitlán revolted and drove the Spanish out but this was merely a stay of execution for the Aztec/Mexica Empire. Cortés retreated to a neighbouring state and then returned with a large force and, most significantly, a small fleet of fighting ships he had built and carried overland. The vessels enabled him to take control of the lake and lay siege to the city. Weakened by lack of food and water, the Spaniards' cannons, cavalry and steel weapons proved decisive and the city was taken and its inhabitants massacred. Tenochtitlán was systematically looted of its innumerable treasures, its buildings and temples were burnt or torn down and thrown into the lake, which was then gradually filled in. Today, it lies beneath a suburb of the metropolis of Mexico City.

Carving of an eagle (facing page) which showed the Mexica/Aztecs where to build the magnificent city that would be destroyed by Cortés's conquistadors (above).

Cortés was not a modest man. He sent many lengthy letters to his King in Spain, detailing the dangers he had faced and the difficulties he had overcome to extend the Spanish Empire on his Monarch's behalf. The map was included to complement the flowing descriptive prose he wrote as he attempted to convey what was clearly his sense of awe and wonder at Tenochtitlán. As Cortés moved from honoured guest to royal jailer to annihilator, the map he sent home is now an important record of a different world, a world that was lost forever when two cultures clashed.

Cortés certainly achieved the fame and riches he sought but, as is so often the case, they proved transient. Envy and distrust dogged his dealings with the Spanish Court and led to him being removed from positions of authority. The wealth that he gained from the Aztec/Mexica treasure clearly passed through his hands quickly as well, because his final letters reveal him begging the King for money to pay off his debts.

Cortés's Palace?

The large and disproportionate sized flag has clearly been added to the map before Cortés sent it home because it is the flag emblem of the King of Spain. Flying high above the city it is surely intended to convey ascendancy and further ingratiate Cortés to his royal patron as his empire is extended. Coincidentally, the eagle that dominates the Spanish Royal Standard was also of significance in Aztec/Mexica legend, for it was an eagle that had led the wandering tribe to this lake on which they built their capital city.

It is likely that the building over which the eagle flag flies was not selected randomly but rather because it was the one Cortés used as a base during the siege and after the conquest. Its proximity to the reservoir and the aqueduct system would have given him control over the water supply to the island city, a vital factor in any siege.

The Aqueduct System

Although surrounded by water the city of Tenochtitlán would have needed a regular and reliable supply of fresh water to ensure the health and well-being of its citizens. The map reveals the sophisticated manner in which this was achieved. A reservoir by the woods fed an aqueduct system that carried fresh water over the lake and into the centre of the city. The technology itself and the organization behind it obviously impressed Cortes, because he wrote about it in detail in an early letter to the King:

> *Along one of these causeways that lead into the city are laid two pipes, constructed of masonry, each of which is two paces in width, and about five feet in height. An abundant supply of excellent water...is conveyed by one of these pipes, and distributed about the city, where it is used by the inhabitants for drink and other purposes.*

The other pipe, in the meantime, is kept empty until the former requires to be cleansed. As the water is necessarily carried over bridges on account of the salt water crossing its route, reservoirs resembling canals are constructed on the bridges, through which the fresh water is conveyed. The whole city is thus served with water, which they carry in canoes through all the streets for sale, taking it from the aqueduct in the following manner: the canoes pass under the bridges on which the reservoirs are placed, when men stationed above fill them with water, for which service they are paid.

The Sacred Precinct

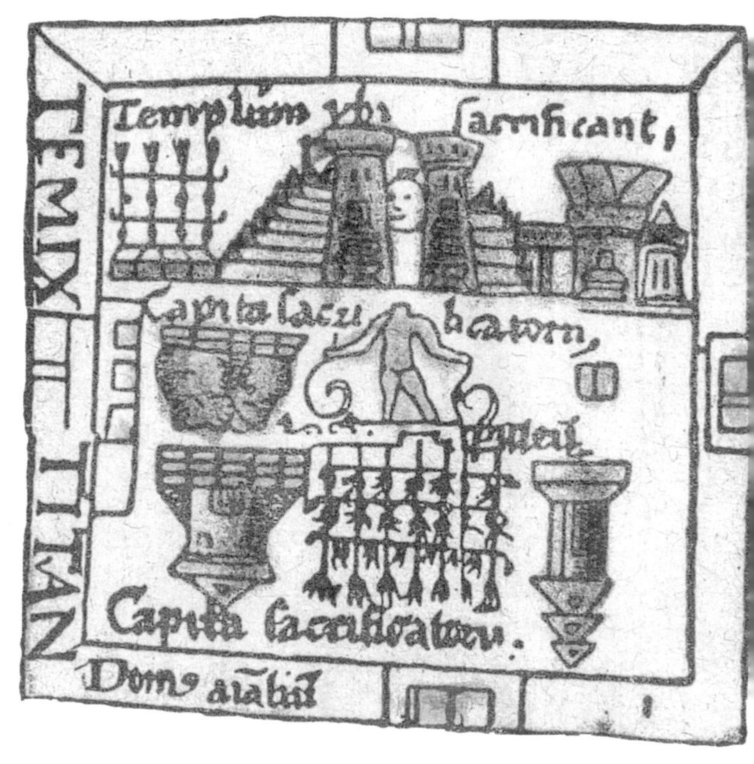

In the Sacred Precinct there were the temples, the most important palaces, the central market, the whorehouse, the ball court providing sport for all as the Coliseum once did in Rome, and there was even a zoo. Among the hustle and bustle would be seen the priests who lived in the temples and who were responsible for ensuring the gods were kept happy and continued to give their blessings to the people of the city and beyond. This goodwill was achieved at a price: regular human sacrifice. To ensure the sun rose again each day captured warriors and slaves were put to death in agonizing ritual ceremonies that involved tearing the heart from the body and returning the blood to the earth. The headless body in the very centre of the square indicates the significance of this aspect of life in Tenochtitlán.

SEPTENTRIO

EQVINOTIALIS

Mare deguine.

Canaga.r.

guinea:~

Mare brasilis:~

Brasilis.

OCCIDENS

Oceanus meridionis.

Mare mg. meridionale:~

Mare argēteū.

Around the Cape

The Coastline of West Africa

By Diego Homen, 1558

Around the Cape

The Coastline of West Africa

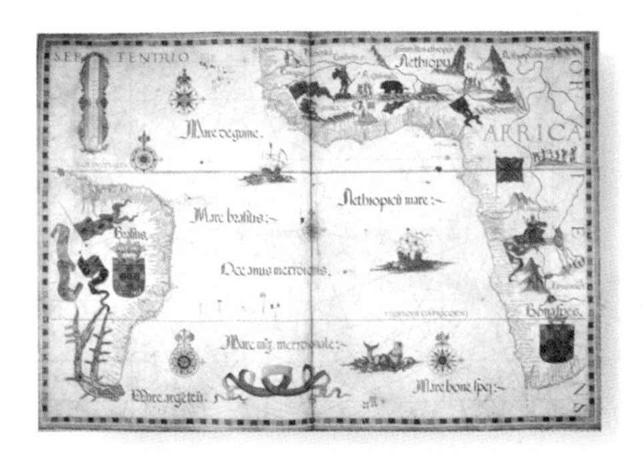

Sibling rivalry is not a new phenomenon. Henry VIII was the father of Mary and Elizabeth Tudor but there was no love lost between his daughters after his death. The map shown here comes from an atlas that was caught up in this bitterness. It was created by Diego Homen (*fl.* 1530–76), a Portuguese cartographer who had fled to England during the reign of Mary I following political exile from his native country. He brought with him much sought after knowledge and skills, for he was from a famous family of explorers and mapmakers. His father had attained the position of Master of the King's Charts in Lisbon. Mary took the opportunity to commission an atlas of the world from Homen and it seems likely that she intended to present it as a gift to her husband, Phillip II of Spain. To show the unity between the two kingdoms resulting from their marriage Mary had ordered that their joint coat of arms be placed in the centre of the map of England. The atlas was completed in 1558 but as Mary died in the same year it passed to her successor, Elizabeth I. On both religious and political grounds Elizabeth loathed Phillip and the story is told that when she was presented with the finished atlas she scratched out the Spanish coat of arms that stood over England. While vandalism inspired by spite is usually condemned, we are fortunate that the atlas did not suffer a worse fate as Elizabeth was renowned for her fiery temper. Homen, perhaps sensing that his association with Mary might not be to his advantage, promptly moved on to Venice where he continued to produce maps of the highest quality.

The pages from the atlas shown here are fascinating because they capture the ongoing interaction between the seafaring nations of Europe and the continent of Africa. The Portuguese sailors, more than those of any other nation, were at the forefront of exploring and mapping the West African coastline. The Mediterranean coast of Africa was well known but the desire to trace the origin of the valuable trade in exotic spices and other goods prompted the Portuguese to sail further and further south along the western coast. Under royal patronage voyage after voyage set sail from Lisbon in the 1400s, each one bringing back not just goods but charts of another section of coastline. Towards the end of the century King John II had special granite pillars taken on each voyage and erected when a previously uncharted area was reached. With what was to become typical European colonial arrogance these pillars were used to 'claim' the land for Portugal, irrespective of who was living there. However, initially it was the information on the charts that was considered of greatest value – so valuable in fact that it was a capital offence to take maps or charts out of Portugal at this time.

In 1487 Bartolomeu Dias sailed south from Portugal with another granite cross. This voyage, however, was to prove particularly significant. At this time European sailors and cartographers believed that the coast of Africa joined on to a massive southern continent that stretched across the bottom of the world. Dias's voyage was to change this belief forever, although it seems to have had more to do with luck than anything else. That is, if one considers being caught up in a ferocious storm that lasted for almost two whole weeks in a small sailing ship as 'luck'! The storm took Dias further and further south. When it finally subsided there was no land within sight so Dias headed north. When his

The thriving port of Lisbon, from an engraving by Theodore de Bry, 1592.

crew did eventually sight land it was on their *port* or western side. The significance of what had happened was quickly grasped by all on board. His crew threatened mutiny unless the ship was turned around and a course set for home. They had become the first Europeans to sail around the southern tip of Africa and enter the Indian Ocean. While uncertain as to what lay beyond the tip it was now clear that Africa did not join on to a southern continent. The map of the world had changed dramatically and irreversibly. Upon his return to Lisbon all maps and charts had to be redrawn to incorporate this new information. Dias is usually credited with changing the name of the southern most tip of Africa from his original 'Cape of Storms' to the 'Cape of Good Hope'. The former might well have been more appropriate for only a few years later he and his ship were 'lost at sea' during another storm in this area.

The Portuguese desperately tried to keep the information secret but knowledge is impossible to restrict to national boundaries and others would soon repeat Dias's voyage around the Cape. It is likely that this particular map from Homen's atlas was among the ones most closely examined by the sea captains of Elizabeth's court when it was published.

Setting up Trade

The Portuguese ships that left Lisbon and other Portuguese ports during the fifteenth century had dual and complementary missions. They were exploring and mapping the African coast but they were also looking for every opportunity to return home with a hold full of goods that would enrich the captain, his crew and, importantly, the sponsors who had financed the voyage. With such small crews the usual way of achieving this was not through confrontation but rather through trading with indigenous people. As in Europe, there were areas that were ruled by powerful monarchs, who controlled local trade. There was no option for the Portuguese but to seek to work with the local ruling elites and to establish ongoing relationships so that future voyages would be aware of the situation they were sailing to. This extract from Homen's map depicts one particularly significant ruler along the West African coast identified with a kingdom south of the River Mamcogo (Congo) in present day Angola. The crescent pennant identifies this as a Muslim kingdom – a convention Homen used throughout the atlas. The manner in which Homen represents this particular ruler was

clearly to convey wealth and riches. The gold on the crown and bracelets is pure gold leaf and still glitters from the map centuries later. Trading with the Europeans further enhanced this local wealth but it needs to be remembered that the most valuable 'goods' traded were slaves. The relatively small-scale dealings of the early Portuguese explorers of this period were to sow the seeds for the massive traffic in human misery that was to dominate the economic and political development of at least four continents for the next 350 years. The ramifications of this continue today.

Mythical Beasts

This map of West Africa was drawn by one of the world's most knowledgeable and leading mapmakers. It is therefore significant that despite his knowledge there is almost no information about Africa away from the coast. This lack of information led to the growth of European myths and legends that assumed the status of fact right into the twentieth century. This extract captures just some of the classic elements of such beliefs. The colourful mountain ranges, the courses of the inland rivers and the locations and activities of the groups of people featured are all illustrative rather than factual. The exquisitely drawn blue beast depicted in the hunting scene seems to be a cross between a rhinoceros, an armadillo and a dinosaur! Like most of the other 'inland' features it served the purpose of filling the space while also making it attractive and interesting to its royal patron. It is not difficult to see how the interior regions of the continent continued to fascinate the European psyche long after the coast had been mapped.

Also of interest is the use of the name Aethiopia (Ethiopia). Here it is used almost as a generic name for the interior region appearing thousands of miles to the west of where the country of the same name is located today.

VIRG
POWHATAN
Held this state & fashion when Capt. Smith was deliuered to him prisoner
MONACANS
MANNAHOACKS
P O W H A T A N
MANGOAGS
CHAWONS
KUSKARAWAOKS
CHESAPEACK BAY
Powhatan
The Fales
Iames towne
Powhatan flu
Poynt comfort
Cape Henry
Cape Charles
Smyths Iles
Russels Iles
Monahassanugh
Rassawek
Monasukapanough
Massinacack
Mowhemcho
Arrohatteck
Appamatuck
Cattachiptico
Passaunkack
Utcustank
Accoqueck
Secobeck
Mahaskahod
Stegara
Shackaconia
Tauxsnitania
Cuttatawomen
Pamacocack
Tauxcenent
Namassingakent
Assaomeck
Quactataugh
Nandsamund
Mattanock
Warraskoyack
Chesapeack
Mortons baye
Kiskiack
Werowocomoco
Accomack
Accohanock
Nause
Nantaquack
Kuskarawaock
Wighcocomoco
THE VIRGINIAN SEA
Scale of Leagues
and halfe
1 5 10
Discouered and Discribed by Captayn John
Grauen by William Hole
38
39

Getting to Know the Neighbours

The Colony of Virginia

By John Smith, 1609

Getting to Know the Neighbours

The Colony of Virginia

John Smith is not a grand sounding name, but the John Smith (1580–1631) who created this first detailed and beautiful map of Virginia was anything but ordinary. He left home at the age of 16 and joined French volunteers fighting alongside the Dutch against the Spanish in Holland. Over the next few years he fought for the Austrians against the Turks in Transylvania, was wounded, captured and sold into slavery before escaping through Poland and Russia. He was only 25 when he returned to England in 1605. He then joined the Virginia Company, which had just been granted a charter by King James I to establish a permanent settlement in Virginia, something that had not proved possible before.

He travelled in one of the three tiny ships that sailed out into the Atlantic in December 1606 carrying just over 100 would-be colonists. When they arrived, after a journey of four months, one of their first acts was to open the sealed box they had carried with them. John Smith learned he was to be one of the seven councillors named to govern the new colony. By late 1608 he had been elected President of the Council and established a tough regime based on his dictum of 'He who does not work, will not eat'.

Despite the anticipation of fortunes to be made, such expeditions were fraught with danger. Lack of food and fresh water, a harsh and unfamiliar climate, inappropriate skills, new diseases, and hostility from the Native Americans were just some of the problems that decimated the expedition's original 104 members within a year of their arrival. Despite these setbacks, Smith enthusiastically set about exploring and mapping the area as a necessary prerequisite to profiting from its riches. This map is a result of the two surveying expeditions he led from the settlement at Jamestown into Chesapeack Bay (original spelling). As the image in the furthest inland corner of the bay shows, they travelled in a shallow draft barge that could be sailed and rowed by his team of about a dozen men. From June through to September 1608 they travelled along both sides of the bay, venturing around every island and up each river they encountered. They recorded the twists and turns of each and Smith noted the natural resources on both land and in the water. What the map clearly reveals is that this was a well-populated region. Smith's key in the top right-hand corner provides the symbols for 'ordinary howses' and, significantly, 'Kings howses'. The colony at Jamestown had been constantly attacked and remained under threat from the indigenous people who were suspicious of their new, uninvited neighbours. Smith realized that unless the colonists could establish friendly links with at least some tribes in the region they were unlikely to survive and he sought out opportunities to meet with local leaders as they completed the survey.

Engraving of John Smith that adorns a map of New England, the area he explored upon his return to America.

Engraving of Pocahontas, princess of the Powhatan tribe, shown here in modern European attire after she arrived in London.

The following year he was injured in an accident at Jamestown when some gunpowder exploded and he returned to England for further treatment. He took his notes and sketches with him and created the map. First published in 1612 the second (Oxford) publication was entitled 'A Map of Virginia: With a Description of the Countrey, the Commodities, People, Government and Religion'. Alongside the map this consisted of Smith's frank and fascinating commentary based on a combination of his notes and memories with a little imagination thrown in to make the whole seem even more exciting, as if the reality was not sufficient! The events of the surveying expeditions, including vivid accounts of disease and death, severe storms and encounters with friendly and hostile tribes are all included.

Smith never returned to Virginia. However, from 1614 onwards he was actively involved in the exploration and early colonization further north in what are now the states of Massachusetts and Maine. Indeed, it is John Smith who is credited with giving the name 'New England' to this area. Despite his numerous adventures and brushes with death Smith lived to the, then, ripe old age of 51. His later years were spent writing books through which he shared the events of his very full life.

Along the Potomac

Smith and his surveying team kept careful records as they sailed and rowed around the bay and up the rivers that drain into it. This extract is taken from his account published in 1612 in England:

> *The fourth river is called Patawomeke & is 6 or 7 miles in breadth. It is navigable 140 miles, & fed as the rest with many sweet rivers and springs, which fall from the bordering hils. These hils many of them are planted, and yeelde no lesse plenty and variety of fruit then the river exceedeth with abundance of fish...*

Today we know the Patawomeke River as the Potomac. Smith's exploration of what he simply described as the 'fourth river' had taken his party through the heart of what is now Washington D.C. This is shown as ✠ on the river (to the right of the extract) and marks the furthest extent of their exploration. They had clearly communicated with the local people for their venture to the higher reaches of the river as shown by ✠ was due to reports of gold and silver to be found there. Despite this information none was discovered and they returned downstream disappointed.

Gyants

Smith's survey of the upper reaches of the bay had established that this tribe was clearly an important one and of an impressive physique. He had sought to establish cordial relations with them both for immediate survival purposes and also for future links. He noted:

> *...the Sasquesahanockes a mighty people... The Sasquesahanockes, inhabit upon the chiefe spring of these 4 [river tributaries]...two daies iourney higher then our Barge could passe for rocks. ...to perswade the Sasquesahanocks to come to visit us, for their language are different: 3. or 4. daies we expected their returne then 60. of these giantlike-people came downe with presents of venison, Tobacco/pipes, Baskets, Targets, Bowes and Arrows,*

Of course, the picture of a 'gyant like' figure on his map added to Smith's kudos and the exotic nature of this 'new' part of the world when viewed in England.

Saved by the Chief's Daughter

Perhaps the most famous of all the stories associated with John Smith arises from this illustration. In his first year in Virginia Smith's hunting party was attacked and his colleagues killed. He was captured and taken to the house of Powhatan, chief of the tribe featured here. He was feasted and entertained but, as he recounted in his book, he remained uncertain of his fate. When his head was placed on what he believed was a ritual stone he thought he was going to be clubbed to death. However, the chief's young daughter intervened on his behalf and he was later released. Her name was Pocahontas and the tale has been recounted ever since. More recent interpretations suggest that Smith had misinterpreted the event and that his life had not been in danger. Rather that the rituals he recounted had meant that he was being accepted into the tribe as an honorary lesser chief!

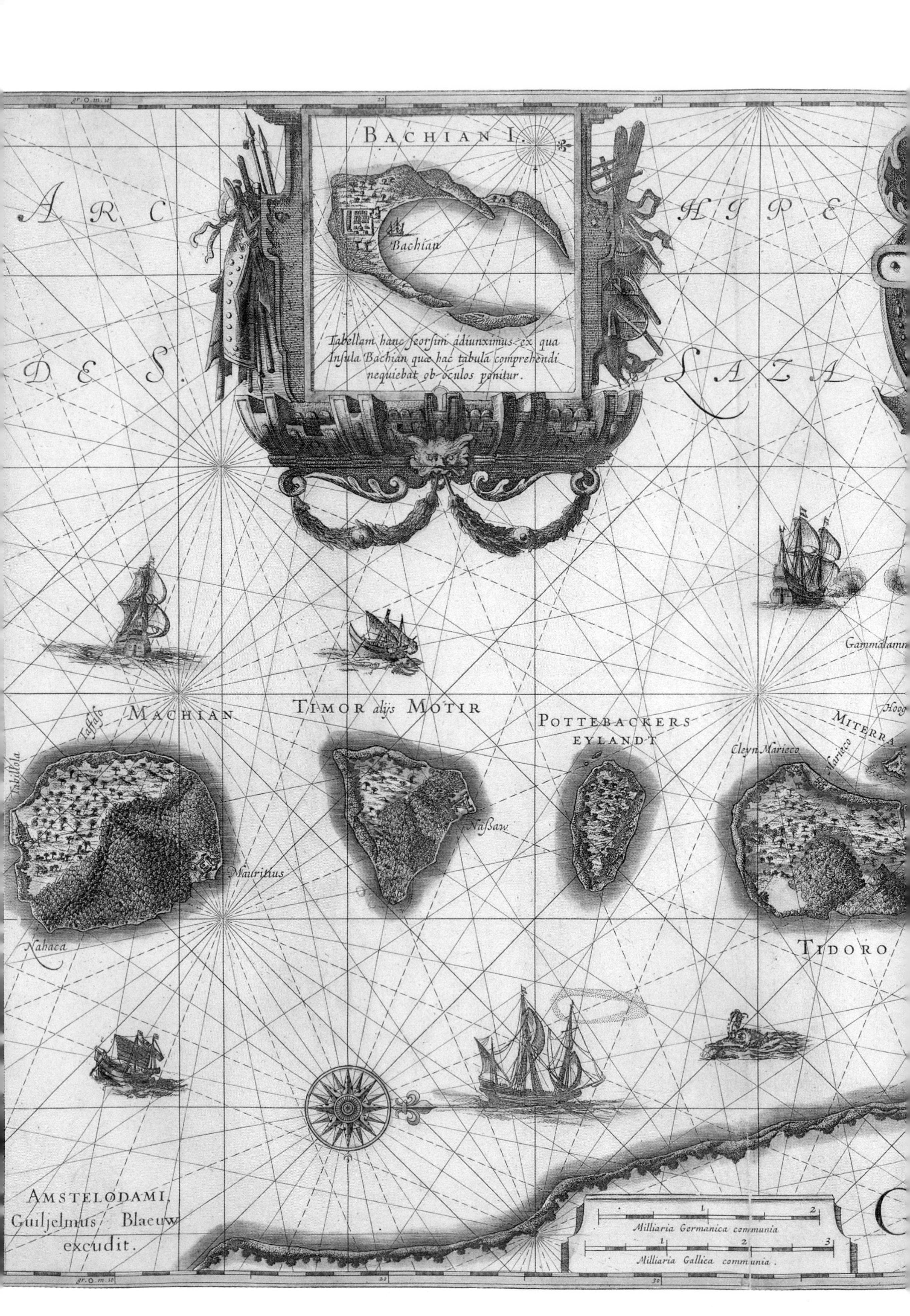
BACHIAN I.
Bachian
Tabellam hanc seorsim adiunximus ex qua
Insula Bachian quæ hac tabula comprehendi
nequiebat ob oculos ponitur.
ARC
DES.
HIPE
LAZA
Gammalamm
MACHIAN
TIMOR alijs MOTIR
POTTEBACKERS
EYLANDT
MITERRA
Cleyn Marieco
Nassaw
Mauritius
Nahaca
TIDORO
AMSTELODAMI,
Guiljelmus Blaeuw
excudit.
Milliaria Germanica communia
Milliaria Gallica communia.

Worth its Weight in Gold

The Spice Islands

By Willem Blaeu, 1630

Worth its Weight in Gold

The Spice Islands

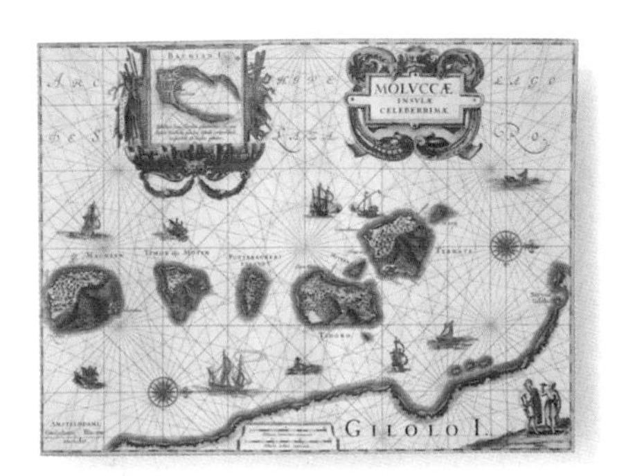

When the words 'treasure' or 'riches' are used they usually conjure up images of buried chests full of gold coins or hoards of precious stones. Today, few people view the jars that make up their spice rack as a treasure trove or consider the tiny seeds, seedpods, leaves and slivers of fragrant bark they contain as 'precious'. Yet this was the case until relatively recently. Such was the value of the trade in spices that countries were prepared to go to war with each other and to commit acts of atrocity in the attempt to dominate this market. What counts as 'precious' has changed through time because no object has intrinsic value; its 'worth' is socially constructed. One of the key factors in determining this value is the scarcity of supply in relation to the demand for it. This is why the spices we use freely cost relatively little today when not that long ago some of them were literally worth their weight in gold.

This map shows what were once known as 'the spice islands' or, more accurately, the Maluku Islands or the Moluccas. The climatic location and distinctive volcanic soils meant that certain spices such as cloves and nutmeg once only grew on certain islands in this group. The use of these and other spices in food is not new. It is known that the Roman, Greek and Chinese civilizations all used them. Until the end of the Middle Ages Europe was dependent on supplies originating from these islands brought overland by Arab merchants. This made them incredibly expensive while also bestowing them with a certain 'exotic' quality.

Portuguese sailors were the first Europeans to sail around the tip of Africa and into

The Dutch East India Company set up their headquarters in Jakarta, Java (above) to protect their interests in the trade of nutmeg and cloves (facing page).

the Indian Ocean. They now had direct access to those islands growing the spices, and Portuguese ships laden with cloves, nutmeg and mace made Lisbon one of the richest, and possibly sweetest smelling, cities in Europe. However, Portugal's monopoly aroused jealousy. Spain, England and the Netherlands all sought to secure a share and, if possible, take control over this source of wealth. Initially it seems that the Portuguese were content to trade fairly amicably but trading outposts quickly became forts as disputes broke out. Portugal's transition from a trading to a colonial presence was rapid and, as ever, it was the native people who suffered. As the other European powers arrived each tried to exploit discontent with the Portuguese and to enlist local allies. Numerous small-scale but bloody conflicts were fought on both land and sea, but it was eventually the Dutch who emerged as the dominant power. More organized than her rivals the Dutch Government issued a charter to the *Vereenigde Landsche Ge-Oktroyeerde Oostindische Compagnie* (VOC) or Dutch East India Company in 1602 asserting a total monopoly of trade in the spices. For the next 200 years this company ruthlessly and systematically exploited the region. To maximize the production of spices such as ginger, cinnamon, cloves, mace and nutmeg, slavery was introduced and those who resisted the military rule were ruthlessly dealt with. When the inhabitants of one island opposed the monopoly genocide was used against its whole population.

The beauty of this map belies what was taking place on the islands featured. It was drawn by the famous Dutch cartographer, Willem Blaeu (1571–1638), in 1630, just as the Dutch began to exert their control across the region. His wonderful artwork captures the way these tiny islands rise almost vertically out of the sea in places, providing just a glimpse of the massive volcanoes that lie below the water. A closer look does highlight the political situation, as a fort can be found on almost every island. These were built, using forced or slave labour, to fulfil several purposes. They provided secure storage and loading facilities for the cargoes of spices destined for Amsterdam and Antwerp. At the same time they acted as a deterrent to rival European powers' attempts to create a presence in the area, while their garrisons also ensured the local population remained subservient.

The colonial period in this part of the world lasted for almost 450 years and the South Moluccas (Maluku Selatan) shown here remained a Dutch colony until 1949 when it became part of the newly formed Indonesia. Unfortunately many tensions remain from the past and peace continues to prove elusive in the islands that are so idyllically portrayed here.

Suppression and Fortification

The island of Ternate was a key one in the Malukun chain. With its neighbour Tidoro, it was the major source of clove production, for the trees grew naturally here. As the Portuguese began the transition from trading partners to colonists it was the powerful Sultan of Ternate, ruler of more than 70 of the surrounding islands who led the resistance. By 1522 the Portuguese had built the fort and then castle at Gammalamme (the name of the volcano that is the island) on the southern tip. Colonization had begun. By the end of the century the Ternatens had joined forces with the Dutch and while this alliance proved successful its longer-term effect was simply to replace one colonial power with another to which the islanders were subjugated.

The fort of Malayo (or Fort Orange) depicted on the bottom left on this extract was an early manifestation of the intentions of the newly formed Dutch East India Company. Built around 1606 its four-square layout replicates the European design of the period. The diagram is detailed and Blaeu may well have consulted the architect's plans, which would have been available in Amsterdam.

The topography of the island is captured very successfully by Blaeu as he cleverly shows one heavily forested side slopping steeply into the sea while behind it a flat cultivated band stretches away. The rocky cliffs reinforce the strategic location of the two forts, each built in positions where the incline to the sea is gentle. Together, they also command the channel between the two important islands of Tidoro and Ternate.

Supremacy at Sea

This map is very succesful in showing how these islands sit in an extensive sea. Indeed there is more sea than land on the map. There is also a clever but not immediately obvious message given to the viewer in the use of the ships portrayed. The Dutch had a very considerable navy at this time and used a large section of it to protect their interests in 'the spice islands'. The patrolling galleons were intended to convey this presence much in the way a policeman on the

beat is intended to do so today. The local dhows with their twin masts and triangular sails are accurately drawn as they ply their trade – under the watchful eye of the Dutch men-of-war.

As this extract demonstrates, the navy was called into action at times. The encounter between the European galleon and the oar-powered fighting dhow is captured in detail. The incredible value of the cargo of the ships heading for Portugal had brought pirates on to the scene and these proved a longer-term problem for the colonists than their European rivals. The smoke adds to the animation of this scene. It may indicate two sets of cannons firing at each other or, more probably considering the origin of the cartographer, the smoke over the dhow is intended to indicate that it has been hit and may be on fire. This would have been a reassuring image for the burghers of Amsterdam who would have purchased copies of this map to hang in their homes and offices.

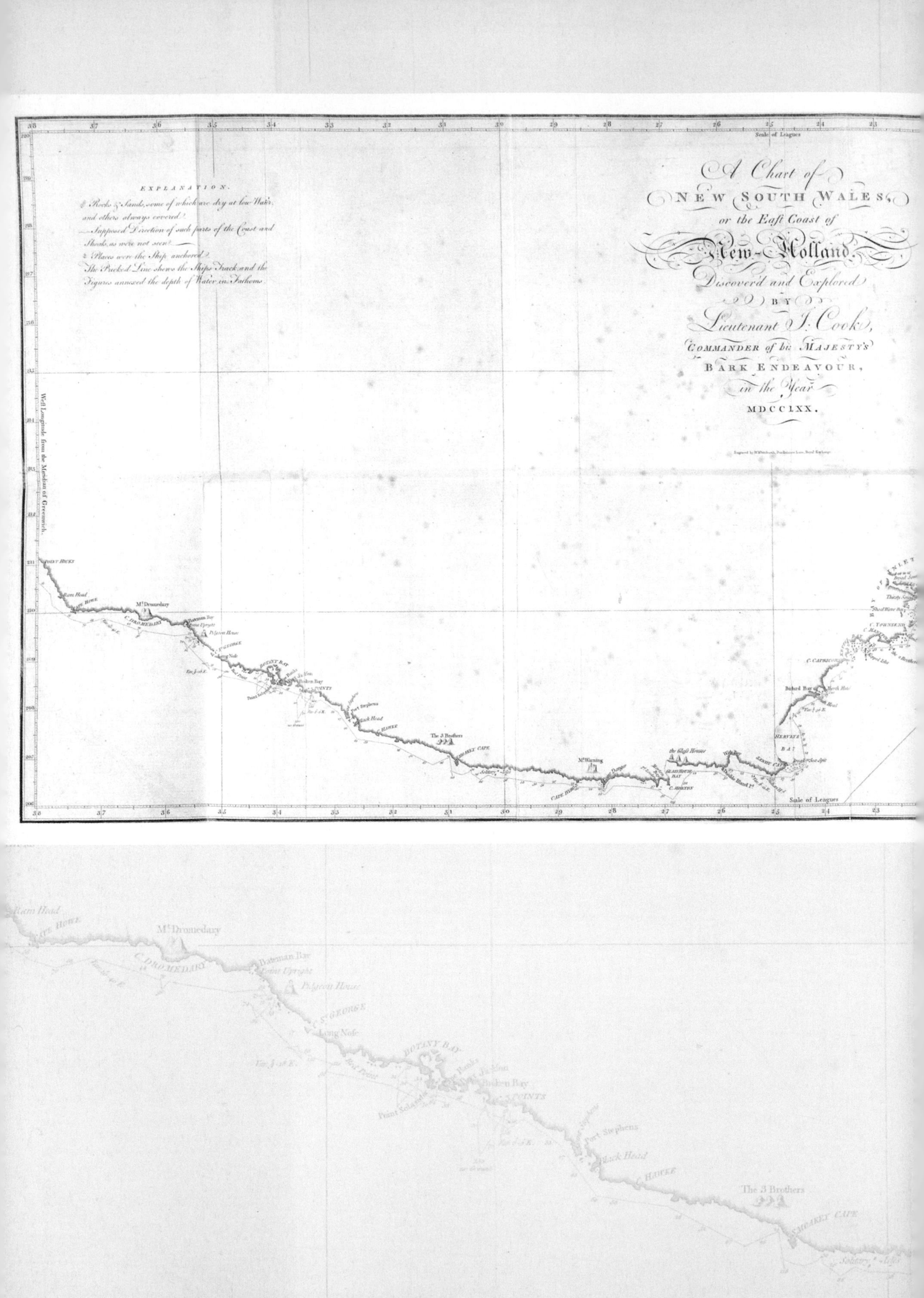

A Chart of
NEW SOUTH WALES,
or the East Coast of
New-Holland
Discover'd and Explored
BY
Lieutenant J. Cook,
COMMANDER of his MAJESTY'S
BARK ENDEAVOUR,
in the Year
MDCCLXX.
EXPLANATION.
Rocks & Sands, some of which are dry at low-Water, and others always covered.
Supposed Direction of such parts of the Coast and Shoals, as were not seen.
Places were the Ship anchored.
The Pricked Line shews the Ships Track, and the Figures annexed the depth of Water in Fathoms.
West Longitude from the Meridian of Greenwich.
Scale of Leagues
Point Hicks
Ram Head
Cape Howe
Mt. Dromedary
C. Dromedary
Bateman Bay
Point Upright
Pidgeon House
C. St. George
Long Nose
Botany Bay
Red Point
Point Solander
Broken Bay
Points
Port Stephens
Black Head
C. Hawke
The 3 Brothers
Smokey Cape
Solitary Isles
Mt. Warning
Cape Byron
the Glass Houses
Glass House Bay
C. Moreton
Hervey's Bay
Sandy Cape
Bustard Bay
C. Capricorn
C. Townsend
Thirsty Sound

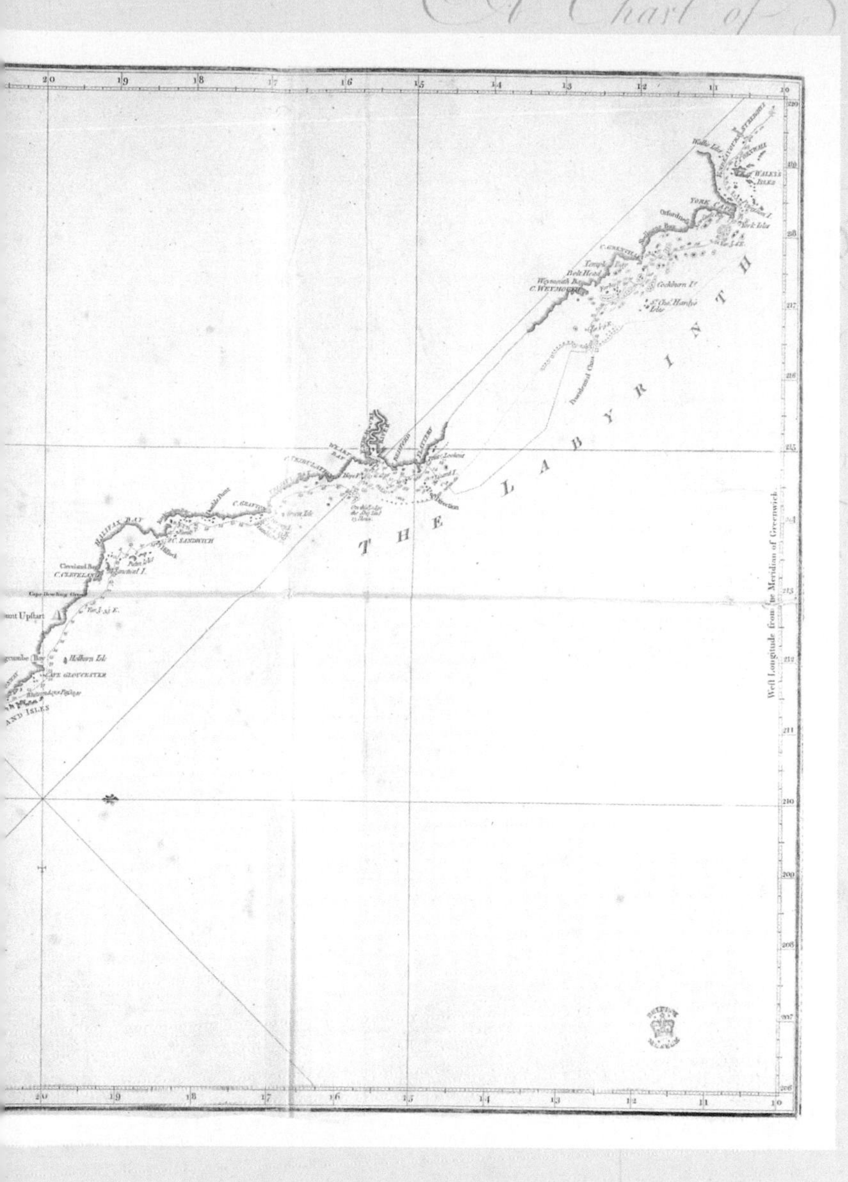

A Job Well Done

The East Coast of Australia

By Captain James Cook RN, 1768

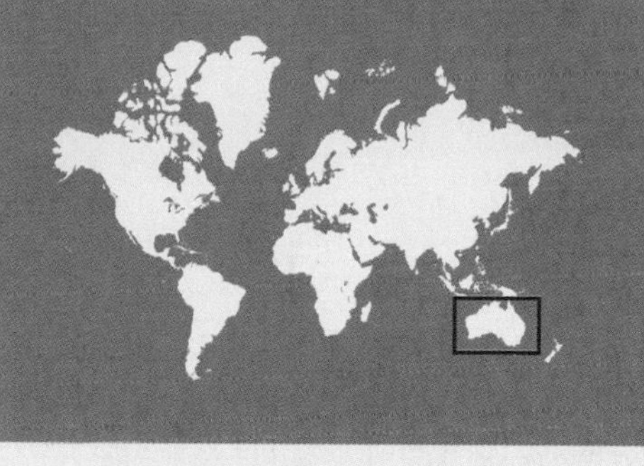

A Job Well Done

The East Coast of Australia

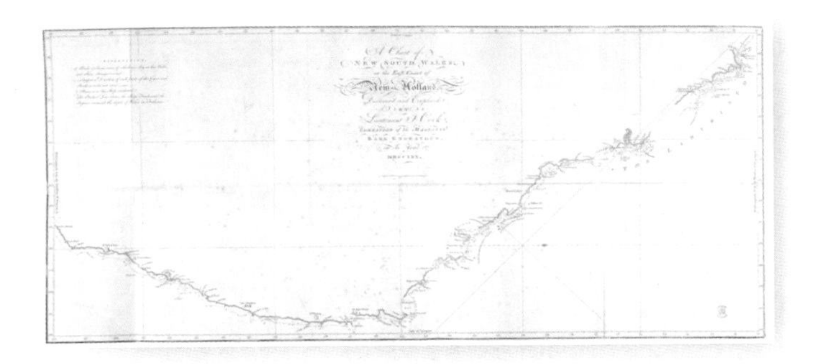

High on the east coast of England where the Yorkshire Moors meet the sea the steep and rocky coastline means there are very few towns or villages. The ancient port of Whitby is the exception and it is from here that a young boy named James Cook (1728–79), coming from a family with no nautical traditions, first went to sea. It was a decision that was to lead to him being responsible for solving one of the last mysteries of the unmapped world. He had a varied and successful career as a naval officer and in 1768 was put in charge of a scientific expedition. Sponsored jointly by the Royal Navy and the Royal Society, its brief was to head to Tahiti to observe the transit of Venus across the face of the sun, a task that was being undertaken from a number of locations. The data obtained would be of scientific interest in its own right but would have practical benefits in terms of assisting navigation. This was a worthy and commendable task, which Cook completed meticulously. But it was at this point that the expedition took a different course.

Together with the orders to observe the transit of Venus, Cook had also been given a second set of orders that were sealed and marked secret. He had been instructed not to open these until the astronomical task had been completed. As he opened the letterbook marked 'Secret Instructions to Lieutenant Cook 30 July 1768' it is not difficult to imagine him feeling a combination of excitement and apprehension. It read, 'there is reason to imagine that a Continent or Land of great extent, may be found to the Southward....

You are to proceed to the Southward in order to make discovery of the Continent'.

The belief that there was a vast southern continent was widely held by the European nations in the eighteenth century. It had been deemed 'a logical necessity' to balance the known continents in the northern hemisphere and had featured on many maps. The reasons for ordering Cook to undertake this mission were also included, '...will redound greatly to the Honour of this Nation as a Maritime Power', but it is clear from the orders that more than this was hoped for as it continued:

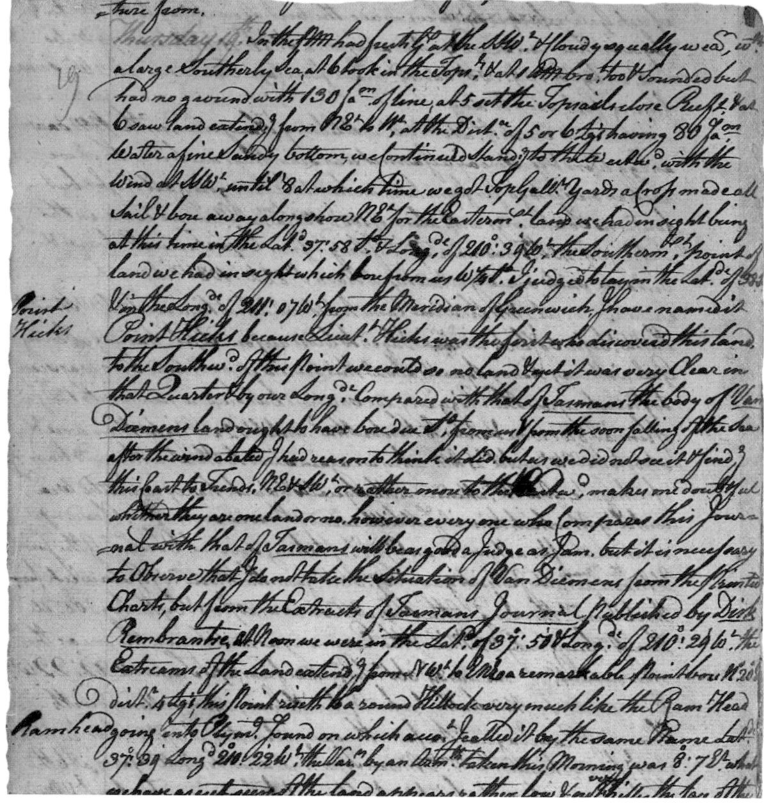

three fathom.

Thursday 19. In the PM had fresh gales at SSW. & Cloudy squally weather, with a large Southerly Sea at 6 oClock in the AM ... brot too & Sounded but had no ground with 130 fathoms of line, at 5 set the Topsails close Reefd. & at 6 saw land extending from NE to W, at the Dist.ce of 5 or 6 Leags having 80 fa. Water a fine Sandy bottom, we continued standing to the Westward with the Wind at SSW until 8 at which time we got Topgallt Yards a cross made all Sail & bore away along shore NE for the Eastermost land we had in sight being at this time in the Lat.de 37°.58' S & Long.de of 210°:39' W. the Southernmost point of land we had in sight which bore from us W ¼ S I judged to lay in the Lat.de of 38° ... and in the Long.de of 211°.07' W. from the Meridian of Greenwich, I have named it Point Hicks because Lieut. Hicks was the first who discovered this land. To the Southward of this point we could see no land & yet it was very Clear in that Quarter by our Long.de Compared with that of Tasmans the body of Van Diemens land ought to have bore due South from us and from the soon falling of the Sea after the wind abated I had reason to think it did, but as we did not see it & find this Coast to trend NE & SW or rather more to the westward makes me doubtfull whether they are one land or no. however every one who compares this Journal with that of Tasmans will be as good a judge as I am. but it is necessary to Observe that I do not take the Situation of Van Diemens from the printed Charts, but from the Extracts of Tasmans Journal published by Dirk Rembrantse. At Noon we were in the Lat.de of 37°.50' & Long.de of 210°.29' W. the Extreams of the land extending from NW to ENE a remarkable point bore N 20 E dist.ce 4 Leags this point rises to a round hillock very much like the Ram Head going into Plymouth Sound on which account I called it by the same Name Lat. 37°.39' Long.de 210°.22' W. the Var.n by an Azim.th taken this Morning was 8°.7' E. what we have as yet seen of the land appears rather low & not very hilly the face of the ...

Point Hicks

Ram head

Following his 'secret instructions' (above), Captain James Cook (facing page), became the first European to survey the coastline of eastern Australia.

> You are also with the Consent of the Natives to take Possession of Convenient Situations in the Country in the Name of the King of Great Britain: Or: if you find the Country uninhabited take Possession for his Majesty by setting up Proper Marks and Inscriptions, as first discoverers and possessors.

Cook followed his orders almost to the letter and after sailing around New Zealand and thus establishing that the two islands were not part of a larger southern continent, he pointed his ship due west. On Thursday, 19 April 1770 the crew sighted the southern tip of the eastern coast of Australia. They were the first Europeans to visit this side of Australia. As the *Endeavour* sailed northwards along the coastline Cook kept a detailed log. He wrote of how they made spasmodic contact with the indigenous population, made notes on the countryside, took soundings and drew navigational charts. In the process, Cook gave English names to the physical features of the bays, rivers, mountains and islands visited and seen. A reference to any modern map of New South Wales will show that the names Cook gave to the topographical features in his log continue in use today.

Seeming to overlook the provisos in his orders relating to 'consent of the natives' and the place being 'uninhabited', Cook's entry of 22 August 1770 reads:

> ...the Eastern Coast from the Lat. Of 38°S. down to this place, I am confident, was never seen or Visited by any European before us.... I now once More hoisted English Colours, and in the Name of His Majesty King George the Third took possession of the whole Eastern coast...by the Name of New

Wales, together with all the Bays, Harbours, Rivers, and Islands, situated upon the said Coast; after which we fired 3 Volleys of small Arms...

The following day, 23 August, the *Endeavour* sailed through the Torres Strait that separates New Guinea from Australia on the journey that would eventually take them home. His mission was complete, although he seemed totally unaware of its significance, concluding that 'the discoverys made in the Voyage are not great', and further that 'the Country itself, so far as we know, doth not produce any one thing that can become an Article in Trade to invite Europeans to fix a settlement upon it'.

Once he returned to England and the expedition's records and charts were collated, the real significance of this voyage became apparent. When the First Fleet arrived less than a decade later the British Empire, quite literally, stretched right around the globe.

Over the next nine years Cook was to undertake two further major expeditions in the Pacific Ocean confirming once and for all that the mythical southern continent was no more than that. He died an untimely death on the third voyage while visiting Tahiti, where he was stabbed to death trying to resolve a dispute over a stolen boat.

Botany Bay, Formerly Stingray Harbour

Botany Bay is an evocative place name and central to the history of Australia. It was here that the 'First Fleet' of 11 ships arrived in January 1788 with the first 700 of the 162,000 convicts that Britain would send to Australia over the next 60 years.

The small anchor on the map shows where the *Endeavour* anchored and Cook's crew then explored further in a smaller boat. One of the purposes was, as ever, to vary the crew's diet and to replenish the ship's food supplies. In the bay the crew of the yawl caught two giant stingrays weighing some 600 pounds (270 kilograms). This prompted Cook to name the bay Stingray Harbour. He then changed it to Botanists Bay as a result of the many different plants found there. Even this got crossed out and his final decision was to give it the name of Botany Bay by which it is known today.

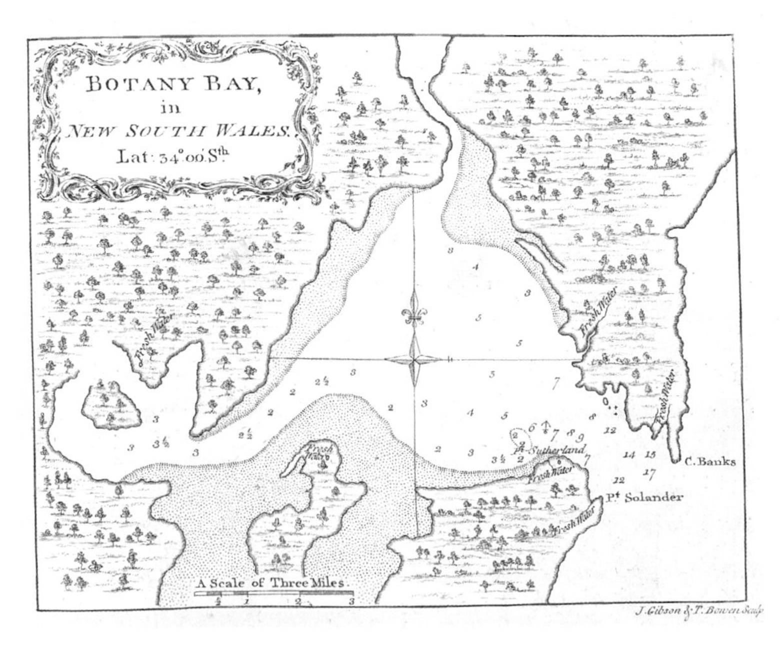

This incident is useful in reminding us of the arbitrary manner in which names that are so familiar to us today acquire their titles and, perhaps more particularly, of the power of the mapmaker to leave his mark on posterity.

Disaster Strikes

It can be easy to forget or overlook just how dangerous these voyages of exploration actually were. Nowhere is this more clearly captured than in the small annotation Cook made on the main map: 'On this ledge the ship laid for 23 hours'. At 11 o'clock on Monday, 11 June 1770, as she made her way up the channel between the Great Barrier Reef and the mainland in what is now North Queensland, the *Endeavour* hit the reef. Initial attempts to float her free by throwing equipment overboard failed. Attempts to drag her free using the smaller boat also failed. Worse, the ship began to take in water and 'heel to starboard'. Cook was not being dramatic when he wrote, 'This was an alarming and, I may say, terrible circumstance, and threatened immediate destruction to us.'

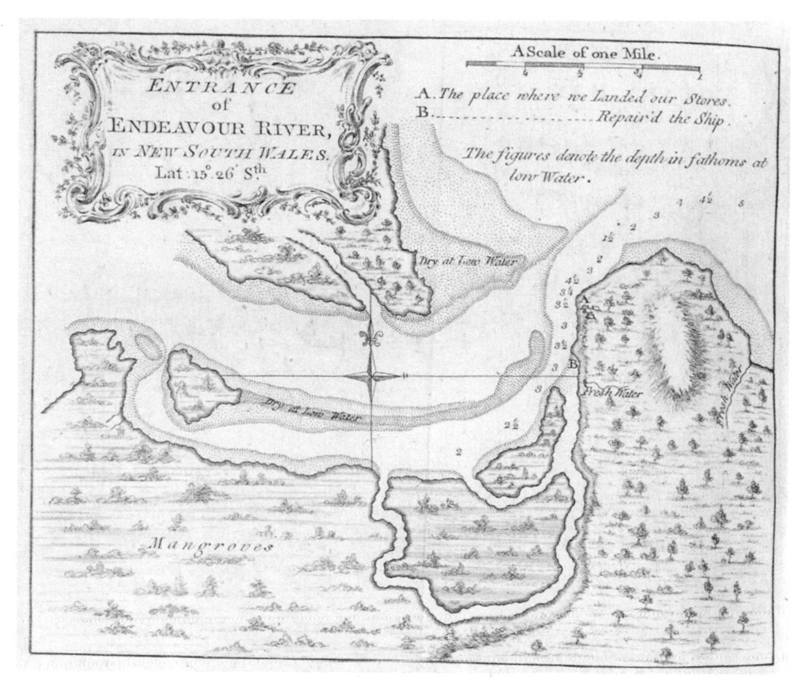

Despite being seriously damaged, the ship was floated free and limped towards the islands nearer the mainland. The origin of their name is captured in his log entry which reads, 'I have named them Hope Islands, because we were always in hopes of being able to reach these Islands'.

As the map shows they did make it into the river which Cook named the *Endeavour*, after the ship. They ran the ship ashore, uncertain as to whether they would be able to make the necessary repairs or not. Cook is generous in paying tribute to his men:

> In justice to the Ship's Company, I must say that no men ever behaved better than they have done on this occasion; animated by the behaviour of every Gentleman on board, every man seem'd to have a just sence of the Danger we were in, and exerted himself to the very utmost.

They were forced to remain ashore here until 4 August making repairs and just surviving, before awaiting the right wind to take them out to sea again and on with their journey. It was while they were marooned here that his log records:

> One of the Men saw an Animal something less than a greyhound; it was of a Mouse Colour, very slender made, and swift of Foot.... It bears no sort of resemblance to any European animal I ever saw...

The sketches of the 'Kanguru' caused tremendous interest when they were seen in London. But the outcome of this voyage could so clearly have been otherwise. There was no one to come to Cook's and his crew's rescue as they struggled to repair their ship. Apart from the Aborigines who watched their efforts, no one else in the world knew of the place where they were.

Chapter Three

Challenging Perceptions

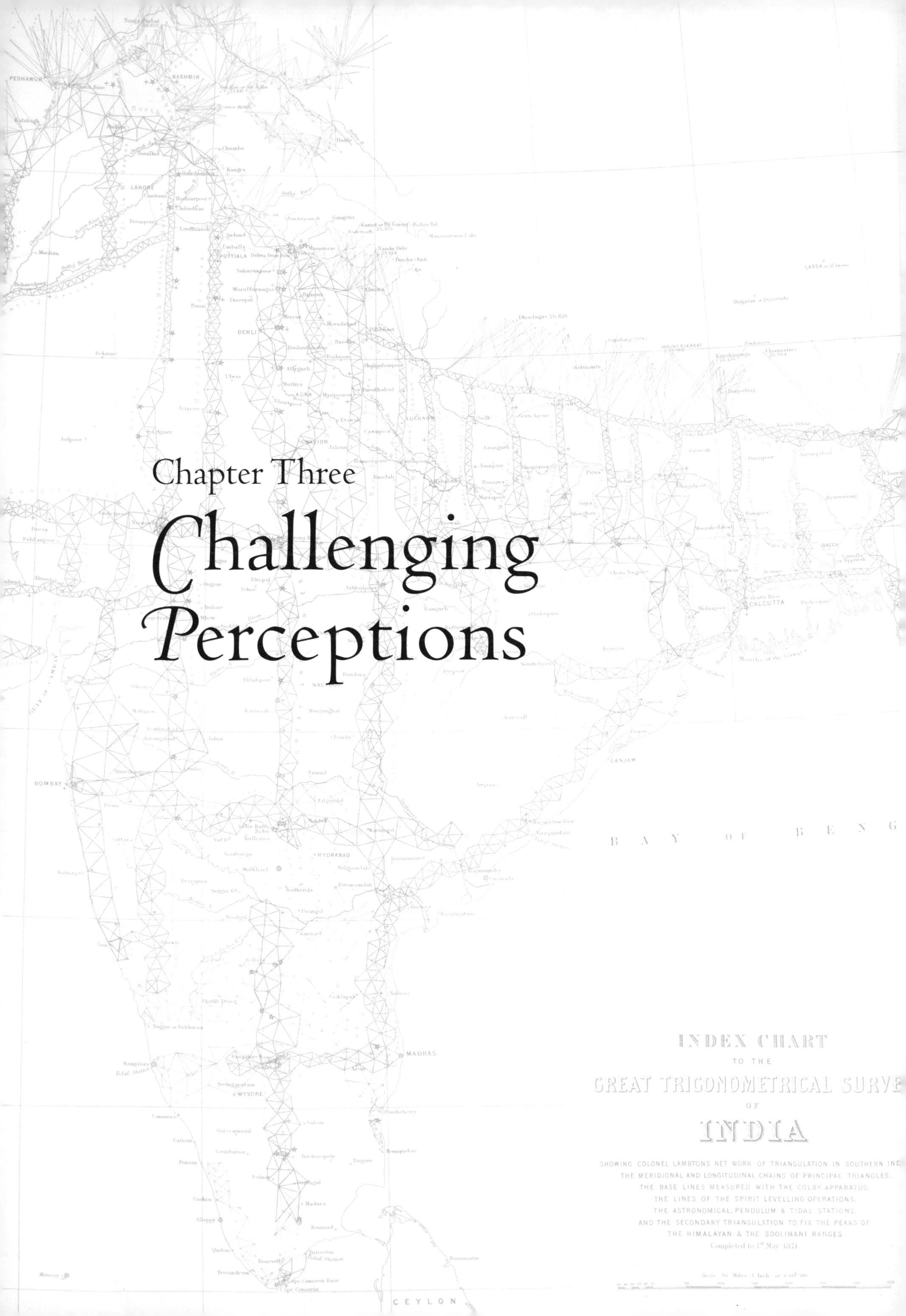

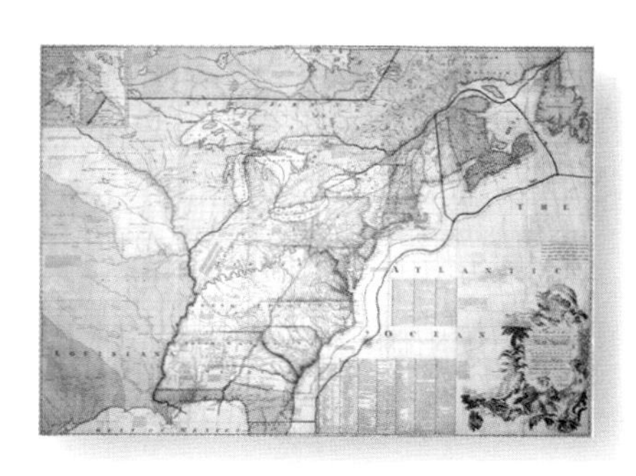

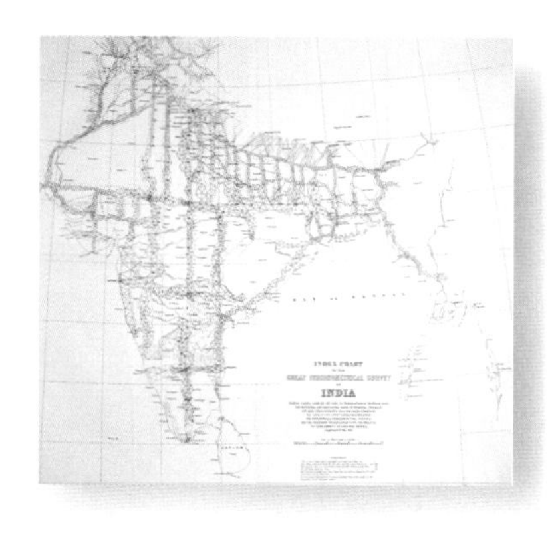

HIGHWAYS OF EMPIRE

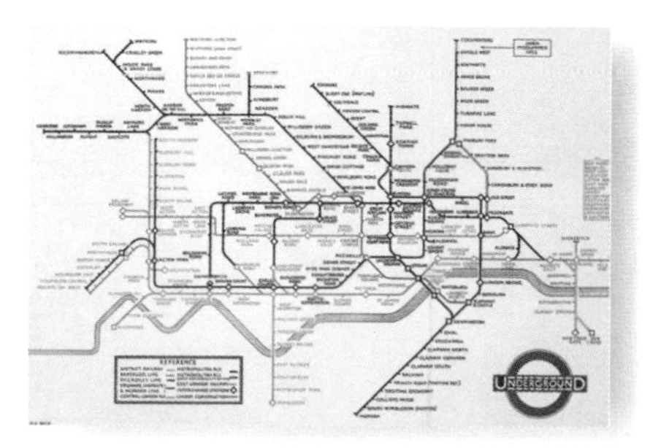

Politicians and advertisers are acutely aware of how the presentation of information affects our 'perception of reality'. Words and images are carefully chosen to influence and persuade. Over time the actual concept of 'the map' has acquired a degree of authority. As their widespread use in news reporting demonstrates a map is an important tool to support an argument or to establish a case. The phrase 'It's on the map' is used in both a literal and figurative sense to denote something that is 'in place' and has a degree of permanency. Such power is clearly open to abuse but it is a situation that is unlikely to change for we are dependent on others to inform us of what is happening beyond our immediate daily lives and to help us make sense of events. While there are more literate people in the world today than at any time in history the power of the visual image has not diminished in any way.

The maps selected for inclusion in this chapter are examples of how this authority is established, used and sometimes taken advantage of. The incredible level of secrecy with which the map the British used in their negotiations with the Americans at the conclusion of the War of Independence (pp. 88–93) was kept for over 100 years is an example of how this authoritative influence can prove enduring. How could a map with a simple red line drawn cause such long-term anxiety and apprehension in what was one of the most powerful nations in the world?

In different contexts the painstaking labours involved in the nineteenth-century surveys of India's landscape (pp. 94–9) and London's poor (pp. 100–105) produced detailed maps whose validity was established at the time and has remained so through to today.

As has been noted the design of the map itself reinforces the message. The location of Jerusalem at the centre of the medieval biblical map in Chapter 1 was an example of this. The 'Empire' map included in this chapter (pp. 106–111) places Great Britain rather than the Holy land at its centre but with the same, deliberate yet subliminal intentions. In both cases the central feature assumes the greatest significance when the whole is viewed. As will be seen when this map is considered in more detail other clever, and not immediately obvious means were used to enhance this importance further.

Charles Booth's survey of wealth and poverty in London in the nineteenth century generated thousands of words. Only the most committed would have read them but the map he assembled from these words was widely consulted and made a major contribution to the debate on what we now call social policy. The information presented on his map was both accessible and immediate. It is a moot point as to whether his survey would have had the impact it did had it simply been presented as a written report.

Maps don't have to deal with major social issues or boundaries between countries to assume the 'authority' that is being explored here. The trust we place in the multi-coloured and many-sized underground map of London (pp. 112–17) is absolute. Anyone making a journey can count off the number of 'stops' on the map as they pass through each station confident that they will correspond and take them to their chosen destination. But is this map really accurate? Geographically the answer is very definitely 'no'. There is a fascinating website (see Exploring Further) which redraws the tube map on screen so that it is geographically correct and relates to the surface features. It quickly becomes obvious that the neat, smooth and symmetrical lines and the evenly spaced stations that are so familiar have little to do with the reality.

Harry Beck designed the famous representational diagram of London's underground railway system 70 years ago.

Can a map ever truly be accurate? New information becomes available all the time, maps have to be updated and earlier ones are relegated to curiosities. One of the maps in this chapter 'owned up' to an error it contained in a very public but imaginatively disarming manner. While maps can charm, fascinate and beguile they are not usually noted as a place where jokes are found. As one of the extracts for the 'Empire' map discusses, the turning of a mistake into 'a joke' was a brave – or desperate – decision made by someone involved in the production process.

The stories associated with the following maps illustrate how they have played a part in affecting the way the world and different parts of it have been presented visually and how this has informed perception. The legacy of 'authority' that the map has is an attribute but one that also needs to be guarded against.

A New Map of Hudsons Bay and Labrador
PART OF BAFFINS BAY
HUDSONS BAY
LABRADOR OR NEW BRITAIN
SOUTH WALES
PART OF HUDSONS OR JAMES'S BAY
CHRISTINAUX or KILISTINOS
NEW FRANCE OR CANADA
LAKE SUPERIOR
EASTERN SIOUX
SIOUX OR NADOUESSIANS
WESTERN SIOUX
MESSESAGUES
LAKE HURON
LAKE MICHIGAN
OUTAGAMIS
MASCOUTENS
LAKE ONTARIO
LAKE ERIE
PENSILVANIA
COUNTRY OF THE PADOUCAS
PANIS
Extensive Meadows full of Buffaloes
VIRGINIA
OHIO
OSAGES
NORTH CAROLINA
CHERAKEES
AKANSAS
CHICASAWS
SOUTH CAROLINA
LOUISIANA
CREEK INDIANS
CHACTAWS
GEORGIA
COUNTRY OF THE CENIS
COUNTRY OF THE APALACHEES
TIMOOQUAS
FLORIDA
GULF OF MEXICO

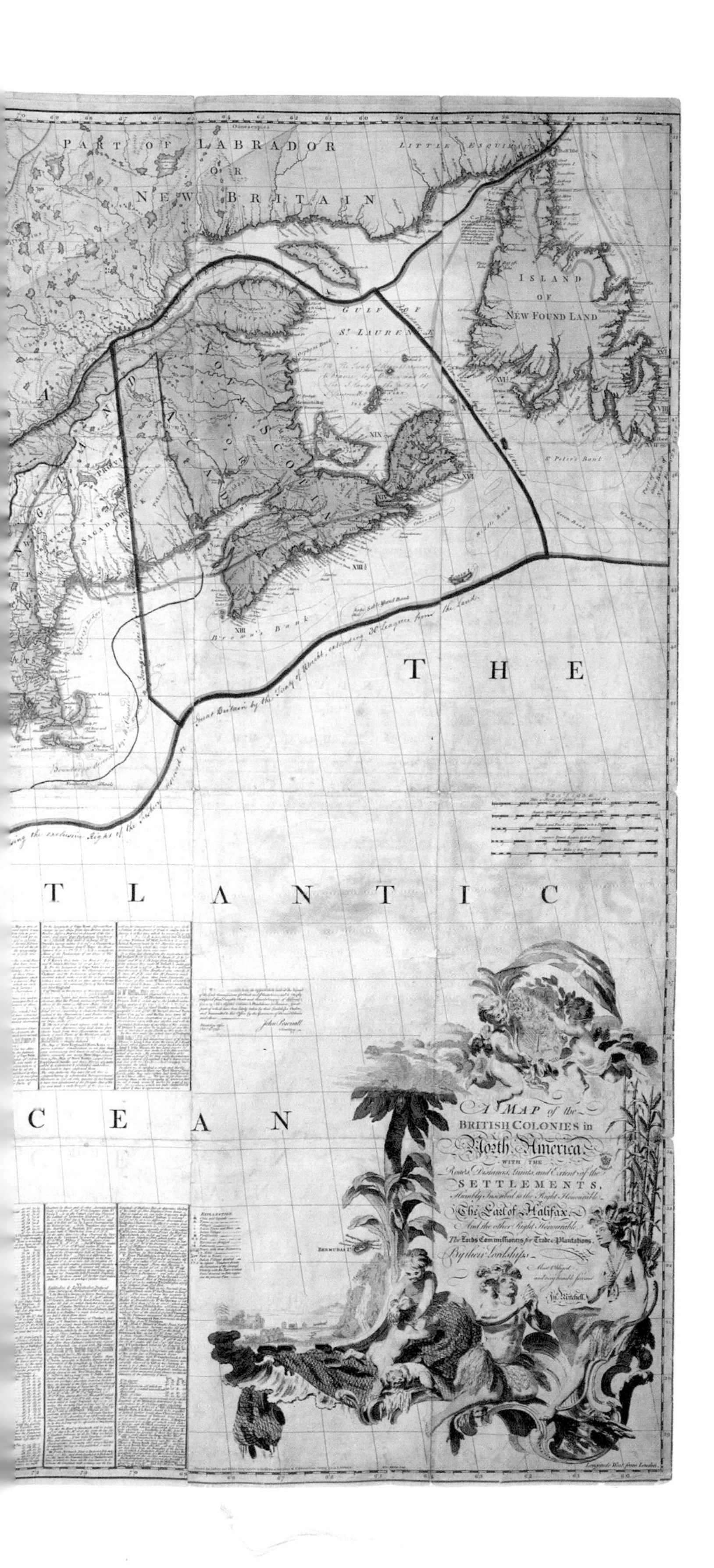

Just a Red Line

The British Colonies in North America

By John Mitchell, 1775

Just a Red Line

The British Colonies in North America

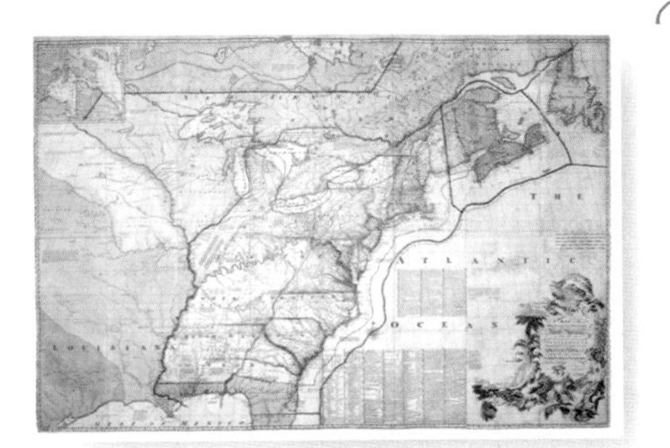

On 3 September 1783 four men signed their seals to a document now known as the Treaty of Paris. As the ink dried on the accompanying signatures of John Adams, Benjamin Franklin, John Jay and David Hartley the existence of the United States of America as an independent nation was formally acknowledged for the first time. The treaty brought to an end the American War of Independence that had begun eight years earlier when colonial militiamen fired on the British Redcoats at Lexington and Concord as they sought to arrest their leaders. During these years the conflict had flared and simmered with neither side able to achieve a decisive victory. Then, in the autumn of 1781 the American army under George Washington combined with its French allies and forced the British army of General Cornwallis to surrender at Yorktown. Although intermittent fighting continued, the war was effectively over.

It might seem strange that the peace negotiations that followed took place in Paris. However, it is often overlooked that the American confrontation had developed into a much wider war involving the major European powers of France, Spain and the Netherlands. The Anglo–American negotiations in Paris proved tortuous. As was customary, both sides had prepared their negotiating positions beforehand to establish what they would seek to gain and what they would be prepared to concede. This map formed part of that process. It was not so much the map itself that was important but more rather the red lines that had been drawn upon it. It was literally 'top secret'.

The thin red line that meanders from the top left corner across the lakes of Superior, Huron, Erie and Ontario, turning sharp right before continuing in a zigzag manner south of the St Lawrence river and then finally turning sharply south to the Atlantic coast forms the familiar modern boundary between the USA and Canada.

Franklin and his fellow negotiators have been credited with what they achieved for the new republic. However, if they had been able to see the other lines that had been drawn on this map they would have realized that they might have achieved even more. The thicker red line that has been drawn around Nova Scotia suggests that the British may have contemplated having to concede this. But this is not the most dramatic possibility.

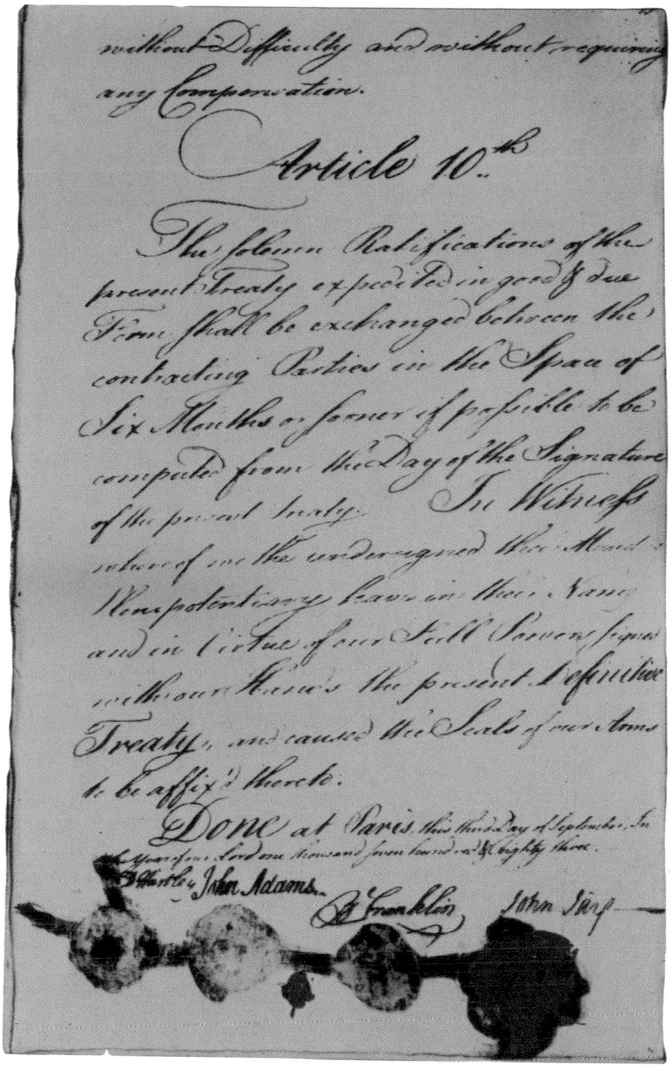
without Difficulty and without requiring any Compensation.

Article 10th

The solemn Ratifications of the present Treaty expedited in good & due Form shall be exchanged between the contracting Parties in the Space of Six Months or sooner if possible to be computed from the Day of the Signature of the present treaty. In Witness whereof we the undersigned their Ministers Plenipotentiary have in their Name and in Virtue of our Full Powers signed with our Hands the present Definitive Treaty, and caused the Seals of our Arms to be affix'd thereto.

Done at Paris, this third Day of September, In the Year of our Lord one thousand seven hundred and Eighty three.

D Hartley John Adams. B Franklin John Jay

Benjamin Franklin (facing page) was one of the American signatories of the Treaty of Paris (above).

Further scrutiny reveals another red line that runs down the middle of the St Lawrence before heading due west to the top corner of Lake Michigan and then south and south-west across to the Mississippi. This seems to indicate the British 'fall back' position or worst-case scenario. If the American negotiators had known this and had been able to exploit the divisions within the British Government at the time, the political division of North America would have been dramatically different. For example, Toronto and Ottawa would today be cities in the USA. Chicago would have been right on the border and Wisconsin a state within Canada! The lines on this map reveal that the British did far better territorially from these negotiations than they had probably expected to.

Despite King George III's protestations on the loss of his beloved colonies, the British knew the terms of the Treaty of Paris could have been far less favourable. Indeed, even after the treaty was signed they remained fearful that the details marked on this map would become known to the Americans and that this would lead to a demand for the boundaries to be re-negotiated. For this reason its very existence was kept secret for the next 100 years. During that time it formed part of King George's private collection and one can only wonder whether he ever looked at it after 1783.

This map is fascinating in many other ways as well. While the USA came into existence in 1783 it needs to be remembered that there were only 13 member states at that time. As the map shows these formed a narrow strip along the eastern seaboard. While each state claimed territory stretching east to the Mississippi this was the limit of their ambition. The area shaded green on the map represents the Spanish colonies which stretched from the Mississippi right across to the Pacific Ocean. As a result of the peace of 1783 Spanish possessions also included the orange shaded area at the bottom of the map around the Gulf coastline and all of Florida. There were many problems still to be resolved as the USA developed into the form we recognize on the map of today.

The red lines on this map act as a reminder of the way chance and fortune have determined the political boundaries that seem so familiar and permanent to us.

The Fate of Florida

As this map extract shows, Florida did not form part of the original union of 13 states. But it did feature in the negotiations that concluded the War of Independence and illustrates the fact that the fighting in America had only been one sphere in a much wider conflict. Alliances based on changing economic interests dominated European politics during the eighteenth century and colonies were transferred between one power and another in the treaties that brought each dispute to a close. Florida was caught up in this process. When England and Spain defeated France in the Seven Years War (1763) Florida was 'taken' from the French and 'given' to Britain. But in the War of Independence France allied with the Americans in their struggle with the British and this time Spain was an ally of France, not Britain. In the peace negotiations of 1783 Florida was 'taken' from the British and 'given' to the Spanish! All this was discussed and agreed over 3,000 miles (4,830 kilometres) away by people who had never been to this state and without any reference to the people who actually lived there. The problems created by this 'map, red pen and ruler' approach invariably created many problems that would have to be resolved at a later date. This was certainly the case with Florida.

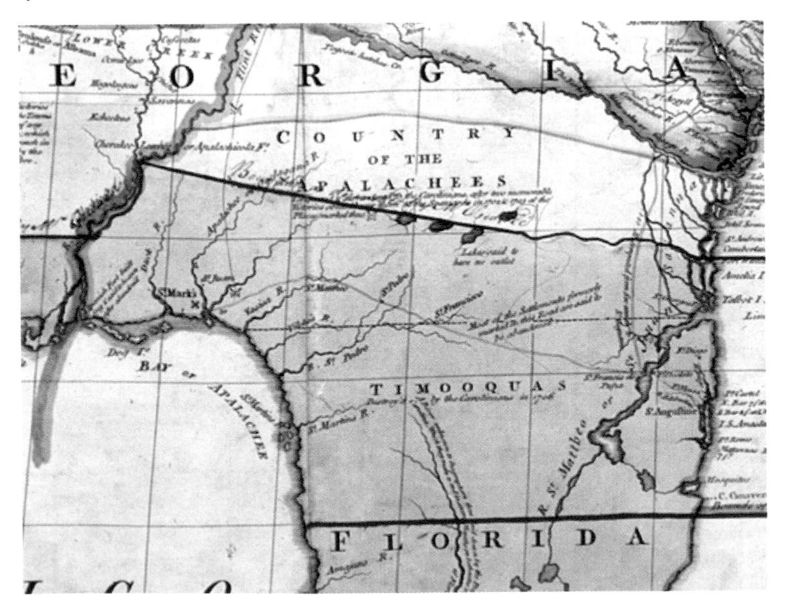

Fishing Rights

The terms of treaties tend to be presented hierarchically with the most important issues appearing at the beginning, and the structure of the Treaty of Paris provides an interesting window into the issues of the day. The third section (article) deals with 'fishing rights' and 'fish curing rights' around the coast of Newfoundland and Nova Scotia. It establishes that 'the people of the United States shall continue to enjoy unmolested the right to take fish of every kind' and:

> *...to have liberty to dry and cure fish in any of the unsettled bays, harbors, and creeks of Nova Scotia, Magdalen Islands, and Labrador, so long as the same shall remain unsettled, but so soon as the same or either of them shall be settled, it shall not be lawful for the said fishermen to dry or cure fish at such settlement without a previous agreement for that purpose with the inhabitants, proprietors, or possessors of the ground...*

This was regarded as successful negotiating by Franklin and his team and was heralded as such. It is very likely that the British also considered this element of the treaty with satisfaction. As the red lines on the map show it seemed that they would have been prepared to concede the whole of Nova Scotia to the new republic.

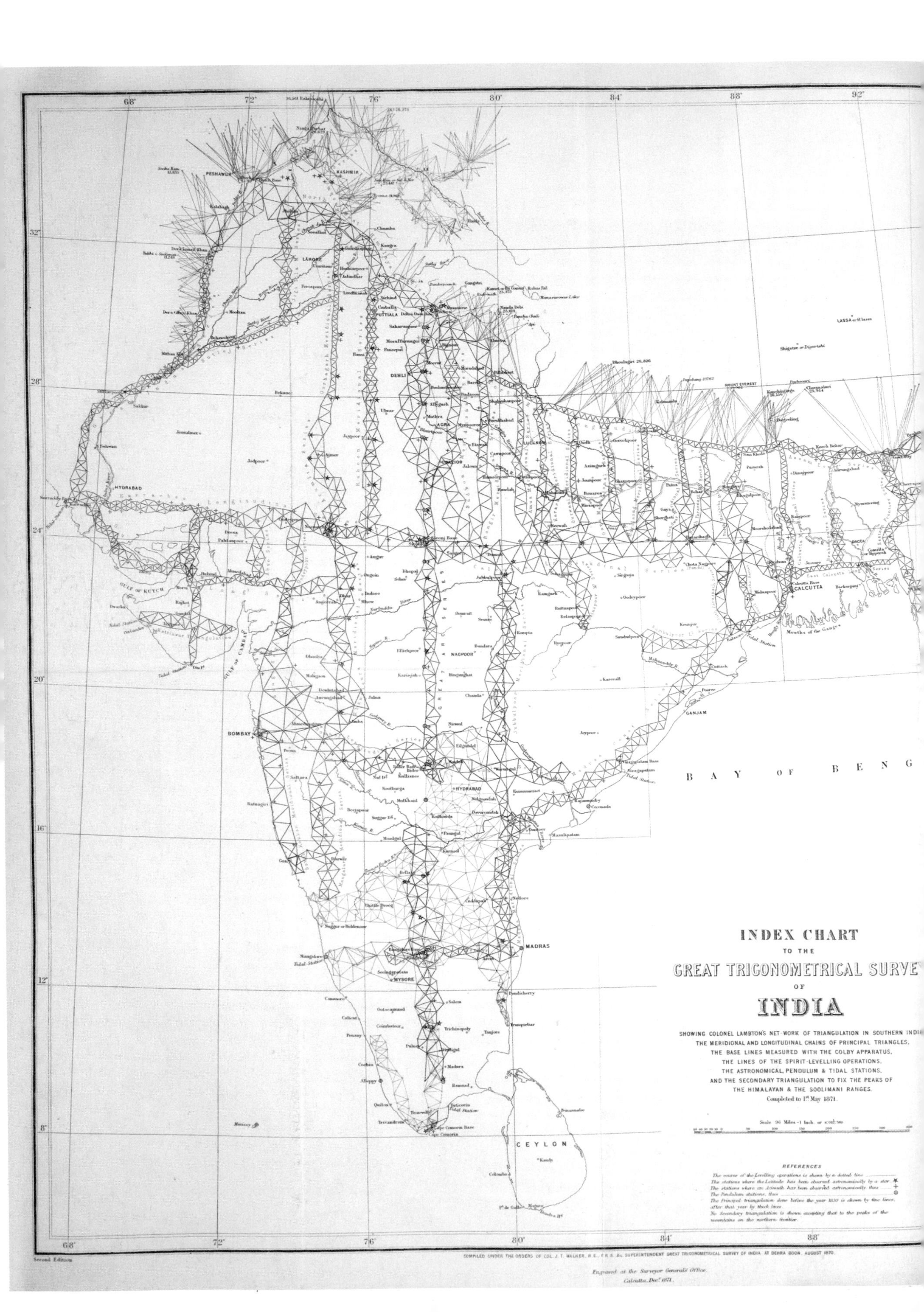
INDEX CHART
TO THE
GREAT TRIGONOMETRICAL SURVEY
OF
INDIA
SHOWING COLONEL LAMBTON'S NET-WORK OF TRIANGULATION IN SOUTHERN INDIA
THE MERIDIONAL AND LONGITUDINAL CHAINS OF PRINCIPAL TRIANGLES,
THE BASE LINES MEASURED WITH THE COLBY APPARATUS,
THE LINES OF THE SPIRIT-LEVELLING OPERATIONS,
THE ASTRONOMICAL, PENDULUM & TIDAL STATIONS,
AND THE SECONDARY TRIANGULATION TO FIX THE PEAKS OF
THE HIMALAYAN & THE SOOLIMANI RANGES.
Completed to 1st May 1871.
REFERENCES
The course of the Levelling operations is shown by a dotted line
The stations where the Latitude has been observed astronomically by a star
The stations where an Azimuth has been observed astronomically thus
The Pendulum stations, thus
The Principal triangulation done before the year 1830 is shown by fine lines, after that year by thick lines.
No Secondary triangulation is shown excepting that to the peaks of the mountains on the northern frontier.
BAY OF BENG
PESHAWUR
KASHMIR
LAHORE
PUTTIALA
DEHLI
AGRA
GWALIOR
LUCKNOW
HYDRABAD
GULF OF KUTCH
GULF OF CAMBAY
NAGPOOR
BOMBAY
HYDRABAD
GANJAM
CALCUTTA
DACCA
MOUNT EVEREST
MADRAS
MYSORE
CEYLON
Mouths of the Ganges
Cape Comorin
68°
72°
76°
80°
84°
88°
92°
32°
28°
24°
20°
16°
12°
8°
Second Edition
COMPILED UNDER THE ORDERS OF COL. J. T. WALKER, R.E., F.R.S. &c. SUPERINTENDENT GREAT TRIGONOMETRICAL SURVEY OF INDIA AT DEHRA DOON, AUGUST 1870.
Engraved at the Surveyor General's Office
Calcutta, Decr 1871.

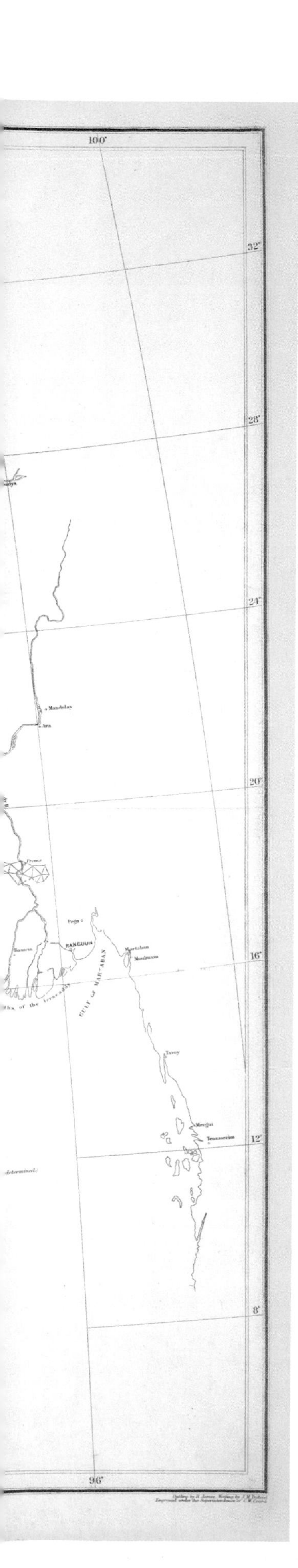

An Obsession With Accuracy

The Great Trigonometrical Survey of India

Published in 1866

An Obsession With Accuracy

The Great Trigonometrical Survey of India

This map came about as a result of British imperialism in India. From its results, communications, defences, taxation districts, etc. were all developed to consolidate imperial control and to exploit the sub-continent's resources. However, this does not detract from the epic nature of the achievement or the commitment, hardship and sacrifice made by all involved in completing such a mammoth mapping undertaking with such an incredible degree of accuracy. The task was begun in 1802 and completed in 1866. Initially it had been prompted by the scientific desire to estimate the circumference of the earth by accurately measuring a section of its north to south curvature (or arc). However, by 1818 the wider significance of creating an accurate and detailed map of the country was recognized and it became known as The Great Trigonometrical Survey of India.

As may be remembered from schoolday maths lessons trigonometry is about triangles and how one can use a combination of knowledge of the length of the sides and the size of the angles to work out the other sides and angles. Despite the advent of satellite technology, this skill remains fundamental to the surveying of height and distance today. The crisscross pattern of small triangles shown on this map show how this process was undertaken. The crucial starting point of such an exercise was the creation of a very accurate baseline. The initial leader of this project, Colonel William Lambton (1753–1823), spent 57 days measuring out a 7½-mile (12.75-kilometre) baseline with a known height above sea level at St Thomas Mount, near Madras. Once satisfied, he used a special theodolite to measure the angle of a visible high point from either end of this line. Using the same formulae used in school maths lessons he then worked out the length of the sides. Each side was then used as the baseline for the construction of the next triangle, and so on and on.

History usually records details of the leaders of such expeditions to the exclusion of the rest of the team. Without the likes of Joshua de Penning and Kaval Lakshmaiah Pantulu in Lambton's original group the whole project may not have got past the first stage. Over the next 60 years the survey initially moved east and then turned north until it eventually reached the Himalayas. It sounds so simple and in one way it was.

However, when one realizes that the sides of the triangles ranged from 30 to 60 miles (48 to 97 kilometres) in length, that the theodolite itself weighed over half a ton and required 12 men to move it, and the fact that it is over 2,000 miles (3,220 kilometres) from the southern tip of India to the Himalayas the enormity of the enterprise can begin to be appreciated. When searing temperatures, monsoons, inhospitable terrain, 'bandits' and wild animals are also added, totally committed project leaders were required. Lambton himself died while travelling between measurement points and his deputy George Everest (1790–1866) took over in 1823. For the next 20 years his column of 4 elephants, 30 horses, 40 camels and 700 labourers on foot slowly made its way northwards completing each side of the hundreds of triangles shown on this map. The topography of India varies dramatically and the vast plains proved particularly problematic as vantage points are necessary for triangulation. The records show that Everest had the unwieldy theodolite hauled to the tops of local buildings, often insensitive to the religious nature of these, and at other times resorted to building 60-foot (18-metre) high stone towers to enable the survey to continue. When his successor as project leader, Andrew Waugh, finally reached the Himalayas and identified Peak XV as the highest in the range, he named it 'Everest' after his predecessor. It is uncertain as to whether Everest ever actually saw the mountain to which his name is now famously linked.

The 'Great' Theodolite was made specially for the Indian survey. Its solidity ensured that the readings were of the highest accuracy.

The mapping of a sub-continent using this mathematically simple but physically demanding method was an incredible achievement and the Great Trigonometrical

Survey compares with any other nineteenth-century scientific accomplishment. Many of its measurements of both height and distance have since been checked using the latest satellite and digital technology. The Great Trigonometrical Survey recorded the height of Peak XV as 29,002 feet in 1849, while today it is officially listed as 29,035 feet (8,850 metres). (In 1999 a re-measurement added another 6 inches to this figure.)

Perhaps a margin of 33 feet can be forgiven because the survey team were forbidden from entering Nepal and had to take their readings from the plains below. However, it is likely that Waugh, Everest and Lambton would not have been satisfied with this error. Such was their obsession with accuracy.

Peak XV

The survey was not allowed to cross the border into Nepal and any attempt to do so was met with hostility. This proved frustrating to Waugh's team as he could clearly see the massive snow-covered mountain range to the north. As this extract shows the course of regular small surveying triangles follows the border and it was from these established heights that

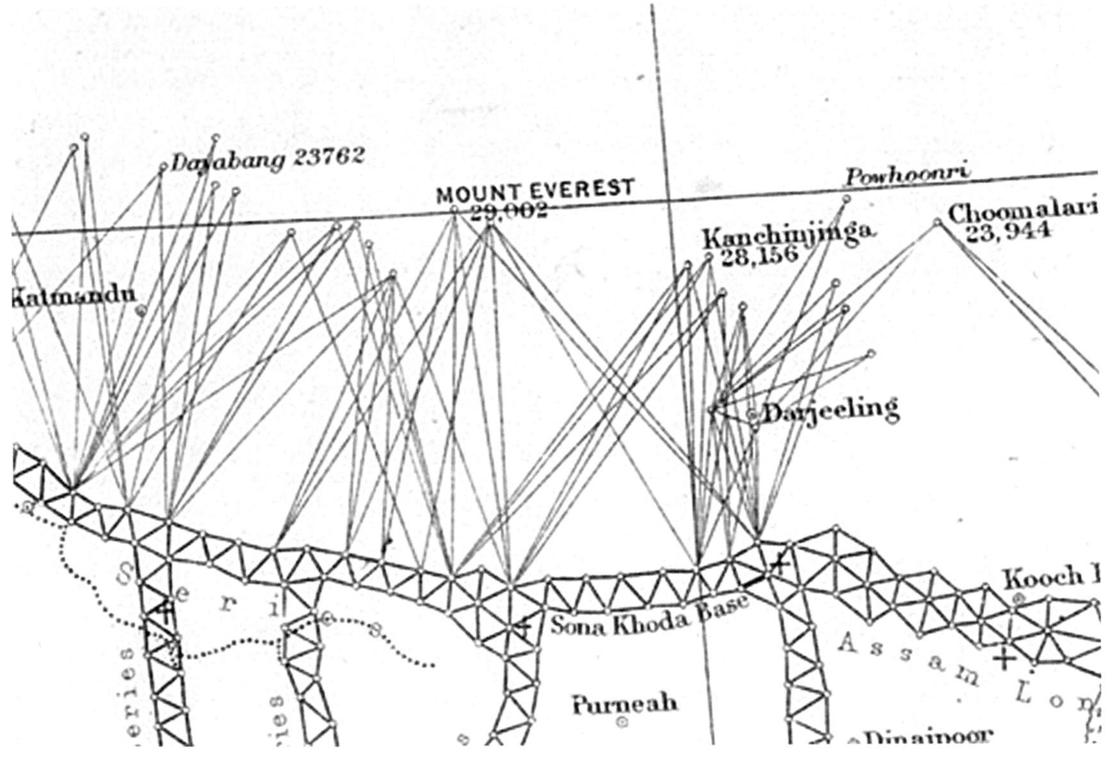

bearings were taken on the distant mountain peaks. It is possible to see that readings were taken from six different positions. When one realizes that the furthest two were over 130 miles (209 kilometres) from each other the time scale involved can begin to be appreciated. One can also see the length of the sides of the triangles increases and this extra distance increases the scope for error. Add to this the fact that the peaks being surveyed were also often lost in the clouds meant a high level of patience was also necessary. Yet, as has already been noted, the accuracy of this first survey of the highest mountain range in the world was quite amazing. There is a story that the readings taken on Peak XV gave it a height of 29,000 feet exactly. Fearing this would look 'too neat' and lead to it being questioned, a further 2 feet was 'found'. Peak XV was now officially the highest mountain in the world.

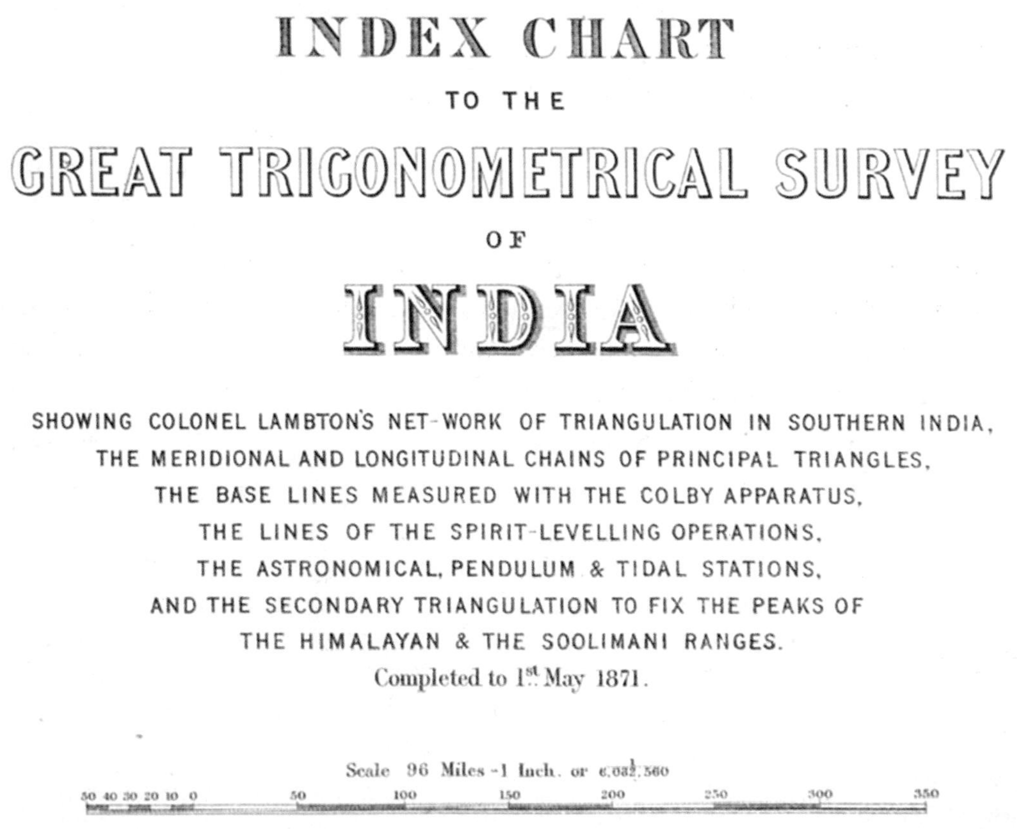

INDEX CHART

TO THE

GREAT TRIGONOMETRICAL SURVEY

OF

INDIA

SHOWING COLONEL LAMBTON'S NET-WORK OF TRIANGULATION IN SOUTHERN INDIA,
THE MERIDIONAL AND LONGITUDINAL CHAINS OF PRINCIPAL TRIANGLES,
THE BASE LINES MEASURED WITH THE COLBY APPARATUS,
THE LINES OF THE SPIRIT-LEVELLING OPERATIONS,
THE ASTRONOMICAL, PENDULUM & TIDAL STATIONS,
AND THE SECONDARY TRIANGULATION TO FIX THE PEAKS OF
THE HIMALAYAN & THE SOOLIMANI RANGES.

Completed to 1st May 1871.

Scale 96 Miles = 1 Inch, or 1/6,082,560

50 40 30 20 10 0 50 100 150 200 250 300 350

In the Interests of Science

There were many pieces of equipment that were vital to the survey as is evidenced by the size of the expedition team as it made its way across the country. The most important piece was the theodolite and the dimensions involved in this project required a special one. It was built by William Cary in London and, due to its weight, needed 12 men to move it. Its readings were so fine that they had to be magnified before they could be read.

Without the 'Great' Theodolite the survey could not have taken place and it so very nearly didn't reach its destination after it was carefully loaded on to a ship in London. On route the ship was captured by the French who were at war with Britain. Luckily, when its purpose was explained to them they repackaged it and sent it on to Lambton in India 'in the interests of science'. Warfare is never civilized but this incident does provide one of those fascinating historical occasions when 'higher' considerations seem to have prevailed over immediate antagonisms.

(IN 12 SHEETS)
ST. SAVIOUR
ST. MARY MAGDALENE
ST. PAUL
ST. MARY
CHRIST CHURCH
ST. MARYLEBONE
ST. LUKE
ST. STEPHEN Westbourne Park
HOLY TRINITY Paddington
ALL SAINTS
ST. JOHN THE EVANGELIST
THE ANNUNCIATION
ST. THOMAS Portman Square
ST. JAMES
ST. PETER Kensington Park Road
ST. MATTHEW Bayswater
CHRIST CHURCH
ST. MARK North Audley St.
ST. GEORGE Hanover Square
ST. GEORGE Campden Hill
THE BASIN (The Round Pond)
SERPENTINE
ST. MARGARET (Out Ward)
ALL SAINTS Knightsbridge
HOLY TRINITY Knightsbridge
ST. MARY ABBOTS Kensington
ST. BARNABAS West Kensington
IMPERIAL INSTITUTE
Imperial Institute Road
ST. STEPHEN
ST. PHILIP
HOLY TRINITY Brompton
ST. SAVIOUR
HOLY TRINITY Upper Chelsea
ST. PAUL Wilton Place
Cromwell Road
EARLS COURT
ST. AUGUSTINE South Kensington
ST. PAUL Onslow Sq.
ST. SIMON Zelotes Upper Chelsea
ST. MICHAEL Chester Square
ST. CUTHBERT Philbeach Gardens
ST. JUDE
ST. PETER Cranley Gardens
ST. MATTHIAS Earls Court
ST. LUKE
ST. JUDE Upper Chelsea
ST. BARNABAS Pimlico
ST. ANDREW
ST. MARY Boltons West Brompton
PARK CHAPEL
CHRIST CHURCH Chelsea
ST. LUKE Redcliffe Square
ST. OSWALD Fulham
ST. JOHN Walham Green
ST. JOHN Chelsea
CHELSEA REACH
8 FURLONGS
1 MILE
THE STREETS ARE COLOURED ACCORDING TO THE GENERAL CONDITION OF THE INHABITANTS, AS UNDER:—
Lowest class. Vicious, semi-criminal.
Very poor, casual. Chronic want.
Poor. 18s. to 21s. a week for a moderate family.
Mixed. Some comfortable, others poor.
Fairly comfortable. Good ordinary earnings.
Middle c… Well-to-…
A combination of colours—as dark blue and black, or pink and red—indicates that the street contains a fair proportion of each of the classes represented by the respecti…

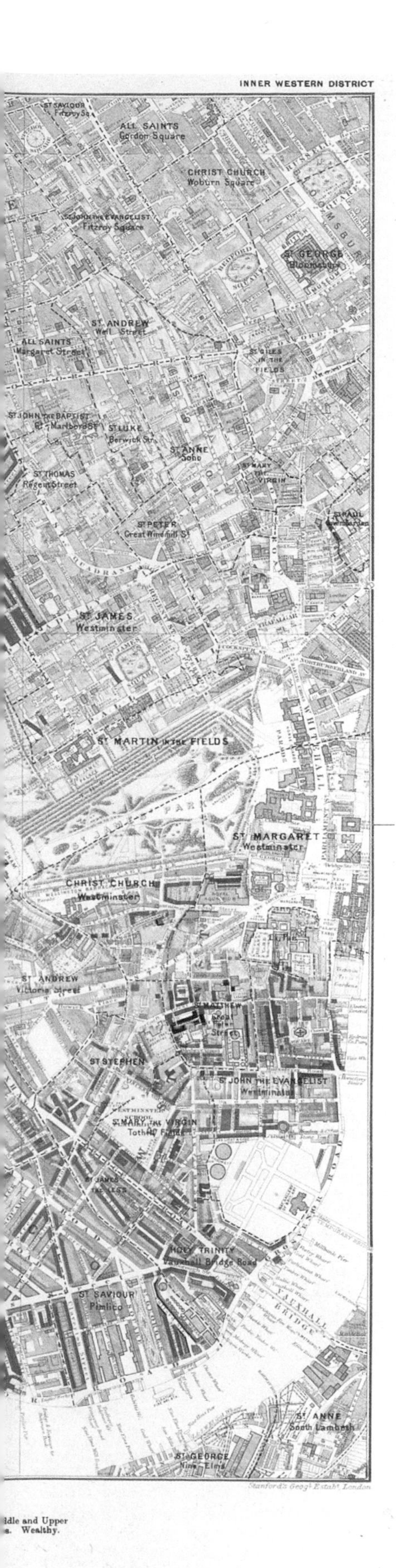

A Grim Tale

Life and Labour of the People in London

By Charles Booth, 1899

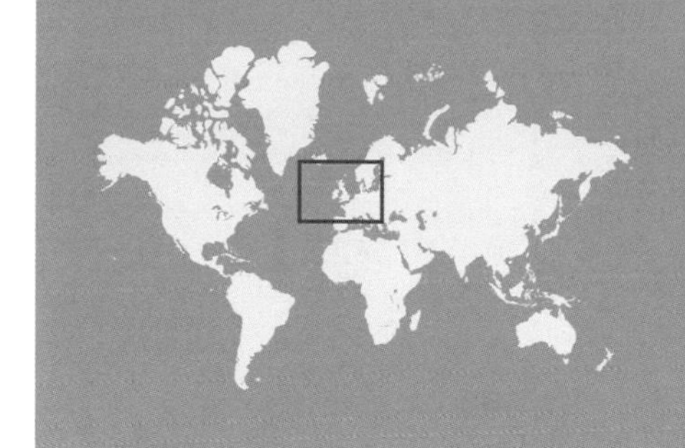

A Grim Tale

Life and Labour of the People in London

Fabulous wealth and abject poverty have existed alongside each other in many societies throughout history. This was certainly the case in Victorian London where the wealth generated by the world's leading commercial centre was far from equally shared. As there were no actual facts and figures available on the extent of these disparities, this led to inflated claims of the problems or simplistic denials that there were any. After Charles Booth's survey and report it was impossible to deny that poverty was a real and severe social issue. Today we take the use of statistics for granted when a case is presented but this is a Victorian legacy, and as such is a relatively new phenomenon. Their fascination with science and its methodology of systematically observing, collecting and collating data was widespread, as was the belief that this data could then be interpreted to provide solutions to 'problems'. Charles Booth confidently applied this methodology to the question of poverty. At a

when very few children were about.
The old inhabitants have all moved into the nearest streets possible. The women into Levy's common lodging houses at the N. end of Whitecross St, in Harrow St, in Marshalsea Rd & in Lombard St. They are Irish cockneys, their bullies are equally in men's lodging houses. The charges of such prostitutes
Maypole Alley
are 1/- or 6d. Sometimes only a pot of beer.
The others have for the most part gone to the Friar St. area esp

Hand-written notes from one of the 450 original survey notebooks.

time when Marx had recently completed writing *Das Kapital* in London and new socialist organizations were being formed on a regular basis, social unrest was an increasing cause of concern. Booth (1840–1916) was not a politician but a wealthy ship owner who seems to have been influenced by his reading of Augustus Compte

Children on the street at the turn of the twentieth century. Charles Booth's survey aimed to find out how many children such as these were living below the poverty line in London.

and his idea that the scientific industrialist would take over the social leadership from church ministers. He can be seen as forming part of that diverse group of wealthy Victorians 'with a social conscience', which included William Morris, John Ruskin and Sydney and Beatrice Webb. The last was actually Booth's wife's cousin, who under her familiar maiden name of Beatrice Potter was involved in collecting the information shown on his map.

In 1886 Booth set out to explore 'the numerical relation which poverty, misery, and depravity bare to regular earnings and comparative comfort, and to describe the general conditions under which each class lives'. His data was collected from three main sources: the records of School Board visitors, the Charity Organisation Society, and from literally walking the streets and making notes. His assistants often accompanied the local policeman as he covered his beat, noting any comments he made on areas or individuals as they walked.

The results were published in 1889 as 'Life and Labour of the People in London'. It had a major impact and showed that far from over-estimating the extent of poverty in London it had been underestimated. Over 30 per cent of the inhabitants of what was probably the richest city in the world were living below the poverty line (which he defined as 18 shillings a week, although the group falling below this figure was further sub divided).

Booth's decision to collate this information and to present it in a visual form was an inspired one, because it made it accessible in a way that a written report could never have done. The original survey covered an area from Notting Hill in the west to Poplar in the east and from Camden Town in the north to Stockwell in the south. However, Booth sought an even more extensive picture and over the next 10 years he expanded his survey area to cover the whole of London and the map shown here is from 1899.

He used seven colours on the map to show the classifications he had devised. As his particular focus was on 'the condition of the working man' he used five colours to identify divisions within this group. For the groups outside he was less specific using red for 'Middle class. Well-to-do' and yellow for 'Upper middle and Upper classes. Wealthy'.

The overall map is made up of 12 sheets and is fascinating at many levels. Viewed in its totality the distribution of each colour makes an immediate impact with the predominance of the yellow of the West End contrasting with the darker shading of the areas to the south and east. However, as the chosen extracts reveal, the closer one looks the more complex the social mix becomes. Then, as now, London was a city of contrasting fortunes living in relatively close proximity.

Booth's publication was one of the first to demonstrate the power of statistics to establish 'the facts'. Although far from perfect by today's standards it remains of huge significance. Most unusually, the original 450 survey notebooks have survived and they contain an enormous amount of detail about the lives of the less well-off Londoners during this period.

Interestingly, one of the recommendations Booth made was for a state-funded old age pension. While initially ridiculed this idea was later enacted by the Liberal government in 1908 and is now regarded by many as the first step in the creation of the welfare state in the UK.

No Change

Anyone who has ever played the game of Monopoly will know that the neighbourhoods of Mayfair and Park Lane constitute the richest and most desirable locations on the board. The vast expanse of yellow on Booth's map indicating 'Upper middle and Upper classes. Wealthy' clearly show that this was the case in the late Victorian period. The carefully laid out gardens that formed Grosvenor and Berkeley Squares, with access strictly limited, formed the heart of this fabulously wealthy sector.

A closer look reveals that the blocks within the grid layout are often yellow on one side and pink representing 'Fairly comfortable. Good ordinary earnings' on the opposite side. These people were the butlers, experienced cooks, grooms, governesses and the many other staff who made it possible for those living on 'the yellow side' to live in style. While they were not well paid they had the advantage of their income being secure and regular. However, it needs to be remembered that only the more prestigious service staff would have been able to afford to 'live out' and the yellow colouring is somewhat misleading. It omits the lesser domestic staff within each household who lived in the tiny attic bedrooms at the top of each house and undertook much of the daily drudgery.

The red shading representing the 'Middle class. Well to do' is found around the edge of the main area with the shops of Oxford Street and New Bond Street forming northern and eastern borders. Shopkeeping, even at the top end of the market, was clearly a middle-class profession.

Paradoxically much, but also little, has changed in this area since Booth undertook his survey. While many, but not all, of the grand houses are now apartments they continue to be inhabited by the very rich. There is a certain irony that the pink areas once inhabited by 'staff' are now highly desirable mews cottages with prices that similarly restrict ownership.

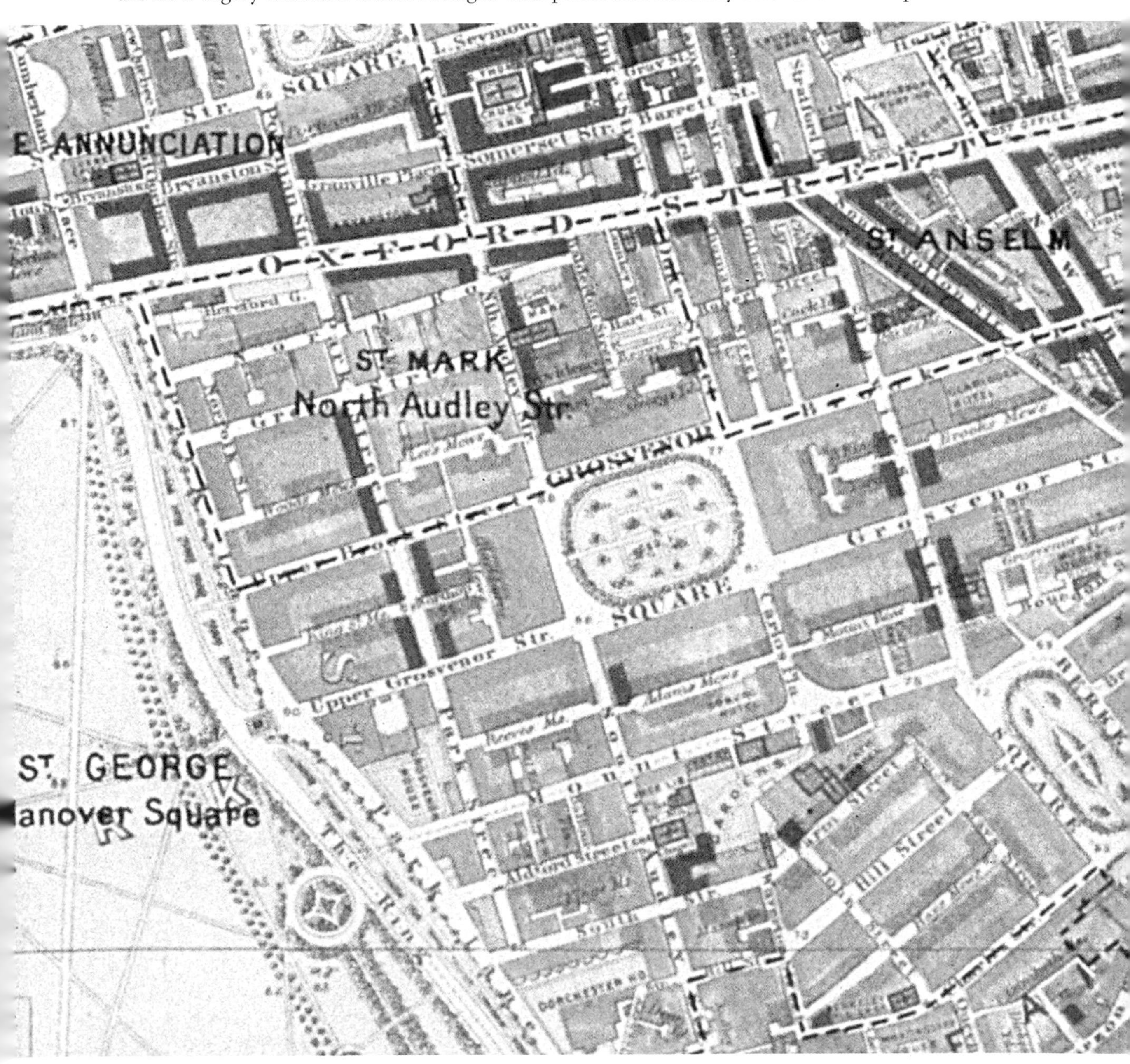

BUY EMPIRE GO
HOME AND O

ISSUED BY THE EMPIRE MARKETING BOARD.

The Artist as Mapmaker

The Highways of Empire

By MacDonald Gill, 1927

The Artist as Mapmaker

The *Highways* of *Empire*

The bold text ensures that the objective of this poster is explicit and unequivocal. Issued in 1927 it exhorts shoppers to give preference to goods from 'The Empire'. As can be seen in the bottom left-hand corner it was issued by the Empire Marketing Board, which existed between 1926 and 1933. Its purpose was quite straightforward: to promote the economic interests of Empire trade. This poster map was specially commissioned to contribute to this.

It is a wonderful example of the power of the map to create a particular desired effect through its actual design. The earth is a sphere. Like any sphere it can be rotated and viewed from any position but none allows the whole to be seen at the same time. This map defies this logic and shows all the continents while maintaining the spherical illusion by using a half round frame.

Every cartographer has to decide what to place in the centre of the map for this is of the greatest significance. It determines the focal point from which all other locations are then related. On this map Britain is placed at the centre of the whole world. Rather subtly, its size has also been increased to boost its importance. The map goes further in enhancing the status of Britain through the inclusion of trade routes, shown as dotted lines, but only those that eventually lead to Britain are marked. A key feature of this map is the numerous little ships – each with their own smoking funnel – eagerly making their way to the heart of the Empire from outposts around the world. The impression is similar to that of a beehive. These ships create a real sense of bustling

Posters commissioned and issued by the Empire Marketing Board between 1927 and 1933 (above and facing page) encourage shoppers to buy goods from the British Empire.

activity and urgency. The subliminal message is that with all this effort and energy being expended to get the goods to *your* shops it would be ungrateful and ungracious, if not unpatriotic, for you, the shopper, not to buy them. It is a case of very clever marketing.

To our eyes the map provides an informative insight into the British Empire at its fullest extent, just before it began to disintegrate. The red shaded areas cover almost a fifth of the land surface and include over 500,000,000 people. It was quite literally true that the 'sun never did set' on the Empire as the earth revolved. Britain was not the sole colonial power: the map also shows the Belgian Congo, French West Africa, Italian Somaliland and the Japanese Empire. As well as the larger areas, it is worth noting the many small islands that are also shaded 'imperial' red for these were very important during this period when steam-driven ships dominated the oceans. The coal and other supplies stored on these islands were crucial to the long voyages undertaken, particularly to Australasia, and control of them was fiercely guarded.

However, empires have always been created and administered for the benefit of the centre at the expense of the periphery. As history records, this empire would unravel very rapidly in the following quarter of a century as one after another the colonies shown here secured their independence. The political map of the world today is vastly different from the one portrayed here.

Alongside its political fascination this map is also significant for its artistic style. The bottom right corner reveals that it was created by MacDonald Gill (1884–1947) who, while not as highly regarded artistically as his brother Eric, had developed a very distinctive style. On this map he combines earlier mapmaking styles and images with his own. The sun and the stars of the night sky, the Latin names, the wind cherubs and the scrolled annotations reflect the classical period. The penguins and polar bears climbing the icebergs of Antarctica, the elephant of India and the kangaroo of Australia are reminiscent of the Renaissance style. But the map's bold colours and strong lines show the impact of the contemporary Art Deco movement. In many ways Gill was one of the last examples of 'the artist as mapmaker' as opposed to the professional cartographer who had emerged as mapmaking became more scientific.

Polar Bears in Antarctica?

Both the North and South Poles are shown with polar bears present. Polar bears, however, are only found at the North Pole. This error was not spotted until the final stages of the map's design and rather than undertake costly changes the mistake was turned into a joke. In what is clearly a late addition, two of the polar bears at the South Pole are given a 'speech bubble'. There are no fancy scroll effects employed nor much attempt to incorporate these bubbles into the overall shape of this part of the world as occurs elsewhere on the map. Indeed, the absence of these features might make one wonder if it was Gill himself who made this addition. There are no rhyming couplets nor clever prose. In one case there is just a simple 'Why are we here? We belong at the North Pole!' Fortunately, this bold, even audacious attempt to deal with the error was accepted and became a distinctive feature of the map's idiosyncratic charm.

The Success of the Suez

There are several severely congested parts of the world's oceans on this map. They include the waters to the south of Australia, the Bay of Bengal and the tip of Africa. None of these areas seem quite as crowded as the Gulf of Aden entrance into the Red Sea through which ships travelled to the Mediterranean via the Suez Canal. This narrow stretch of water was particularly crucial to the trade with Asia and East Africa and at times the queues for passage through the canal itself would have resembled the busiest motorway on a Bank Holiday weekend! The only alternative was the lengthy and often stormy route around the Cape of Good Hope. This dependence on the Suez Canal would lead Britain into ill-judged military action some 30 years later.

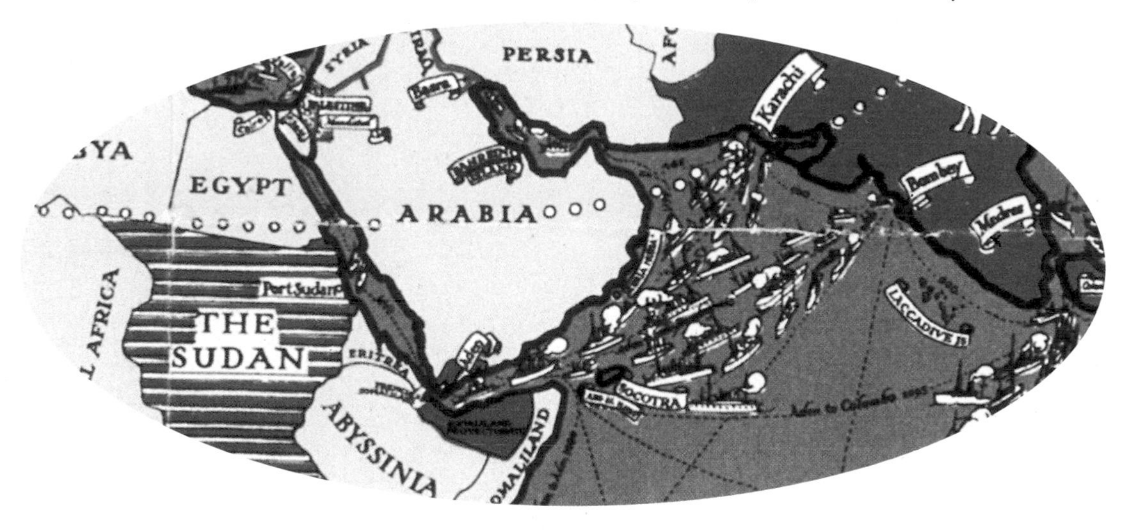

Wind, Steam or Air?

The fascinating insight this map provides into the past, the present of 1927, and the future means of international trade is a further interesting, yet almost certainly unintentional feature of this map. Although the smoking steamships dominate the trade routes, the fully rigged sailing ships in the Atlantic Ocean hark back to a previous era. They would still have been in use but in ever decreasing numbers. What is unlikely is that Gill would have anticipated the significance the form of transport he included near the top of the map would assume over the next half-century. The biplane shown here had limited carrying capacity and even more limited range when compared to any of the merchant ships. When the map was created in 1927 planes were barely into their second decade but their advantages of speed and directness were already appreciated. The race was on in the industrial nations to take the technology forward so that this potential could be realized.

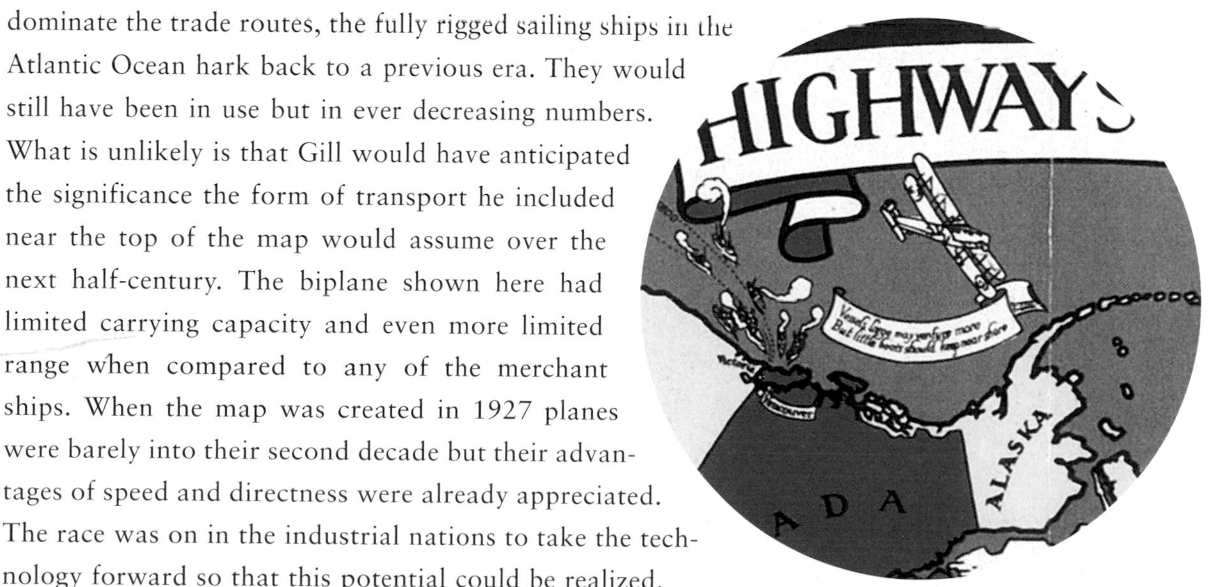

WATFORD
RICKMANSWORTH
CROXLEY GREEN
MOOR PARK & SANDY LODGE
NORTHWOOD
PINNER
NORTH HARROW
HARROW ON THE HILL
WATFORD JUNCTION
WATFORD (HIGH STREET)
BUSHEY AND OXHEY
CARPENDERS PARK
HATCH END FOR PINNER
HEADSTONE LANE
HARROW & WEALDSTONE
KENTON
STANMORE
CANONS PARK
KINGSBURY
NEASDEN
PRESTON ROAD
EDGWARE
BURNT OAK (WATLING)
COLINDALE
BRENT
UXBRIDGE
ICKENHAM
RUISLIP MANOR
RAYNERS LANE
WEST HARROW
NORTHWICK PARK
NORTH WEMBLEY
WEMBLEY PARK
DOLLIS HILL
HILLINGDON
RUISLIP
EASTCOTE
WEMBLEY FOR SUDBURY
STONEBRIDGE PARK
WILLESDEN GREEN
KILBURN & BRONDESBURY
HARLESDEN
WEST HAMPSTEAD
BELSIZE PARK
SOUTH HARROW
WILLESDEN JUNCTION
FINCHLEY ROAD
SUDBURY HILL
KENSAL GREEN
SWISS COTTAGE
SUDBURY TOWN
QUEENS PARK
ALPERTON
KILBURN PARK
MARLBORO ROAD
MAIDA VALE
PARK ROYAL
ST. JOHNS WOOD
WARWICK AVENUE
MARYLEBONE
NORTH EALING
LATIMER ROAD
WESTBOURNE PARK
ROYAL OAK
EALING BROADWAY
WEST ACTON
EAST ACTON
LADBROKE GROVE
BISHOPS ROAD
EDGWARE ROAD
BAKER STREET
PADDINGTON
BAYSWATER
PRAED STREET
NORTH ACTON
WOOD LANE
NOTTING HILL GATE
LANCASTER GATE
BOND STREET
EALING COMMON
SHEPHERDS BUSH
UXBRIDGE ROAD
HOLLAND PARK
QUEENS ROAD
MARBLE ARCH
OXFORD CIRCUS
SOUTH EALING
NORTHFIELDS
GOLDHAWK ROAD
PICCADILLY
BOSTON MANOR
ACTON TOWN
DOVER STREET
OSTERLEY
ADDISON ROAD
HIGH STREET KENSINGTON
HYDE PARK CORNER
SOUTH ACTON
KNIGHTSBRIDGE
HOUNSLOW EAST
HAMMERSMITH
GLOUCESTER ROAD
BROMPTON ROAD
CHISWICK PARK
BARONS COURT
EARLS COURT
HOUNSLOW CENTRAL
STAMFORD BROOK
HOUNSLOW WEST
VICTORIA
GUNNERSBURY
TURNHAM GREEN
RAVENSCOURT PARK
WEST KENSINGTON
SOUTH KENSINGTON
SLOANE SQUARE
WEST BROMPTON
KEW GARDENS
WALHAM GREEN
RICHMOND
PARSONS GREEN
PUTNEY BRIDGE
REFERENCE
DISTRICT RAILWAY
BAKERLOO LINE
PICCADILLY LINE
EDGWARE, HIGHGATE & MORDEN LINE
CENTRAL LONDON RLY.
METROPOLITAN RLY.
METROPOLITAN RLY. (GREAT NORTHERN & CITY SECTION)
EAST LONDON RAILWAY
INTERCHANGE STATIONS
UNDER CONSTRUCTION
EAST PUTNEY
SOUTHFIELDS
WIMBLEDON PARK
WIMBLEDON
H.C. BECK

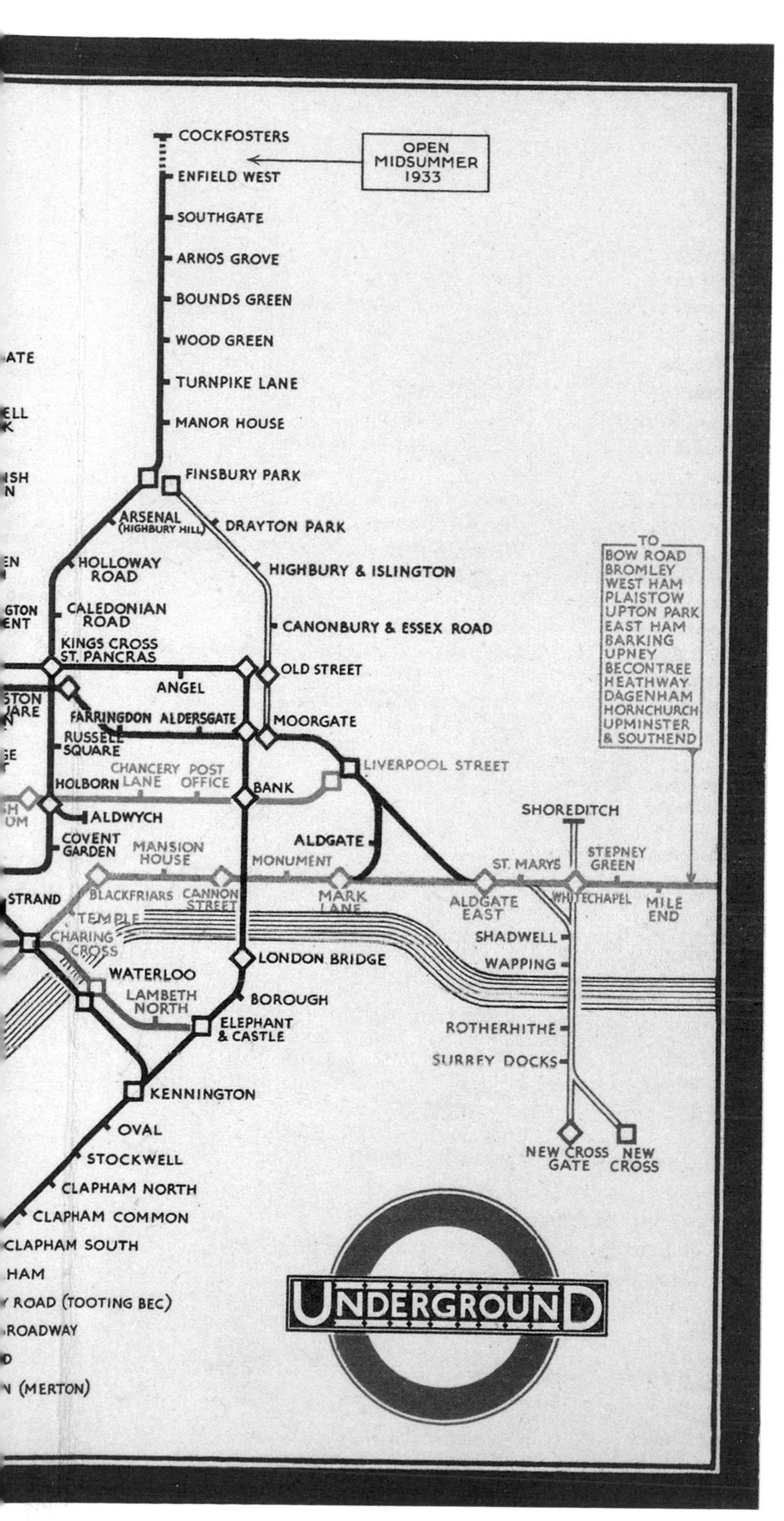

Ten Guineas Well Spent

London Underground

By Harry Beck, 1933

Ten Guineas Well Spent

London Underground

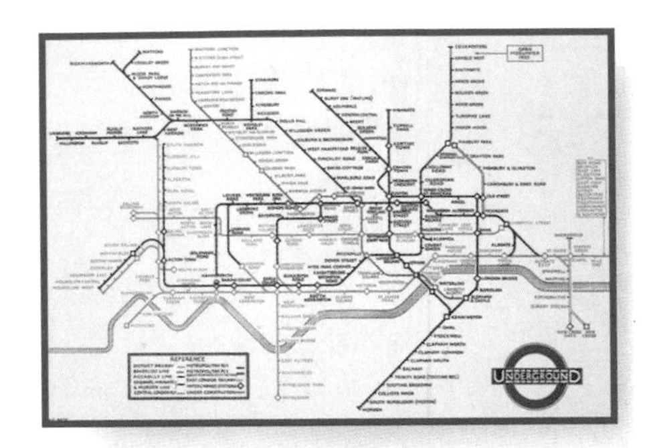

Some images are so familiar to the eye that it is difficult to imagine them being otherwise. The 'Tube' map of London is an example of this. Barely glanced at by Londoners as they make their daily journeys it is a 'must have' necessity for any visitor to the city. While small changes are regularly made as a new line is completed or a station renamed the overall design remains reassuringly familiar. This was not always the case and the pursuit of a map that was informative, clear and attractive proved elusive. For over 70 years after the world's first underground railway carried passengers between Paddington (Bishop's Road as it was called then) and Farrington Street stations in 1863 there were numerous attempts to create such a map. The traditional approach in which the underground lines were superimposed upon maps showing above ground features produced cluttered and confused results. As the system expanded the problem became more acute.

When a fresh perspective was offered it came from an unexpected quarter. The late 1920s and 1930s were particularly difficult times with unemployment reaching record levels in many countries. Harry Beck (1903–1974) was one of many electrical draughtsmen working in the Signal Engineer's Department of London Transport. His contract was a temporary one and, like so many others, he was 'laid off' when the economic crash of 1931 hit London. Possibly to occupy the time he turned his mind to how London's underground railway map might be improved. He applied his experience as an electrical draughtsman working with circuit diagrams to the problem. When the economy improved and he was 'taken on' again by London Transport he showed his design to his colleagues and they encouraged him to send it to the Underground's Publicity Office. The absence of all geographical information except the River Thames and the manner in which he ignored the rules of scale by compressing the outer areas and expanding the central one was branded as 'too revolutionary'. Nevertheless, he re-submitted it a year later and it met with a more favourable response. In 1933 the Publicity Department cautiously issued a trial version and invited the travelling public to give feedback. The response was overwhelming – they loved it!

Beck was paid the princely sum of 10 guineas for his creation, but did receive a promotion. Except for a brief interlude, he was responsible for the development of the underground system map until 1960 when, in less than harmonious circumstances, he moved to an academic career in design. Under his direction the map had new lines added, stations renamed and deleted, colours changed and other minor modifications made. But the overall 'look and feel' of the original 1933 map shown here was maintained. The inspiration Beck had taken from electrical wiring circuit diagrams had enabled him to achieve the designer's elusive goal: functionality coupled with pleasing aesthetics. His success led to many awards and his design has been emulated by urban transport systems around the world.

Poster for the London Underground, which was issued around 1930, just a few years before Beck's map went into circulation.

The 1933 map is at once recognizable but also 'not quite right' to anyone who knows the current London Underground map.

For example, a closer examination reveals some unfamiliar station names – Addison Road, the interconnecting station on branch lines of the District and Metropolitan Lines, Dover Street on the Piccadilly Line and British Museum on the Central Line, to mention but a few. What happened to these?

And the colours? Where is the bright yellow of the Circle Line at the heart of the system? Why isn't the Central Line bright red? Why are all the station names in colour? What are all those diamonds where we are used to seeing circles? In the Reference box, why is the Northern Line known as the Edgware, Highgate and Morden Line?

The answers to these questions can only be found when one realizes that the map guide to London's underground railway system has, and always will be, in a state of evolution. The changes and modifications that have been made, and will be made in the future will be driven by the same factors that challenged Harry Beck when he first picked up his drawing pens: fundamentally, to allow users of the tube to plan and make their journeys with ease and confidence.

Such is the iconic status that Beck's design has attained that it is inconceivable that a completely new one will be introduced. Each year millions of paper copies are issued free of charge and an interactive internet version can now be accessed world-wide. It is also printed on souvenir mugs, hats, T-shirts and even underwear, which London's millions of tourists take home with them to mark their visit.

The 10-guinea fee paid to Harry Beck for the map certainly proved to be a very sound investment.

Ghost Stations

There are nearly 40 'ghost stations' on today's underground. These are the stations that once formed part of the system but for different reasons are no more. Yet today's trains race through many of these and despite the absence of lights that have long been switched off it is said that the discerning eye can still make out the dark platforms. One such is the British Museum station whose white tiles can be seen as the Central Line trains roar through the tunnel between Holborn and Tottenham Court Road. It was closed in 1933 just after Beck's map was first published. Some stations, such as City Road and Down Street station had already gone. The latter was located between Hyde Park Corner and Dover Street (now Green Park) on the Piccadilly line and was deemed 'surplus to requirements' in 1932. It achieved legendary status long after its closure when it was used by Winston Churchill and his war cabinet during the Blitz. Aldwych Station became the latest 'ghost' when closed in 1994. Unlike many of its peers it is likely to remain familiar to future generations for it has been 'preserved' and is used by TV and film companies for location scenes.

There are two further names in this extract that do not exist on the present-day map that you may enjoy trying to identify.

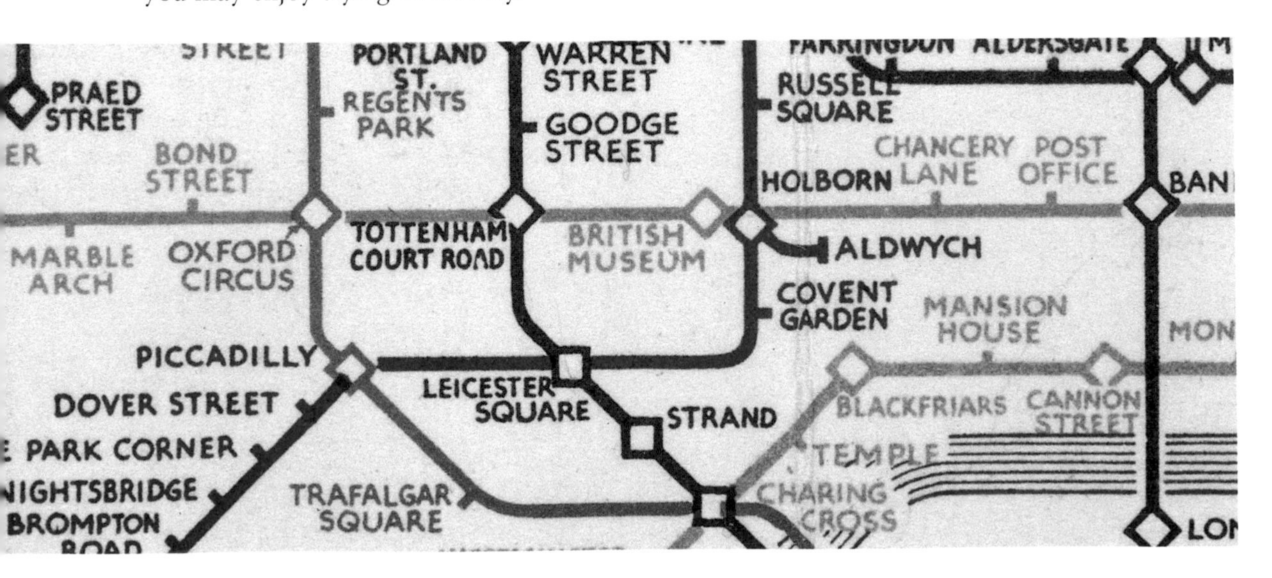

A Familiar Symbol

The logo is an integral and distinctive element of London's system identifying its stations, trains and staff. Like the map itself the version that appears on Beck's map is familiar but subtly different from the one used today. While the roundel and the blue band are constant, the presentation of the text has gone through a number of stylistic changes. The 1933 one shown here reflects the artistic influences of the period. Of greater significance perhaps is that one finds no references at all to the word 'tube' on maps of this period. The name had arisen from the way the tunnels were bored and constructed. However, a senior manager of the 1910 period apparently disliked this term and it was banned from all publications and painted over where it existed. It was even chiselled out from actual station names, including that of the Aldwych mentioned previously. To Londoners the underground has always been 'the tube' and this reality was officially acknowledged in 1990. The word now features prominently on everything linked with the underground system.

Quicker to Walk

Harry Beck identified that tube passengers really didn't need to know, or care, what surface features they were travelling under. Indeed, the inclusion of the River Thames on his map and its presence ever since is as much to do with artistic considerations as practical ones. The absence of any geographical information, however, can cause difficulties for the less familiar user, as Bill Bryson's recent book *Notes From A Small Island* has recorded so amusingly. In 1933 a visitor using Beck's map to travel from Bank to Mansion House would probably have taken a train to Liverpool Street, transferred on to the Metropolitan Line to Mark Lane then transferred again to the District Line and travelled three further stops to his/her destination. They would have emerged barely 250 yards down Queen Victoria Street from where they had first descended underground. It is likely that many travellers have been, and will continue to be, bemused by such experiences.

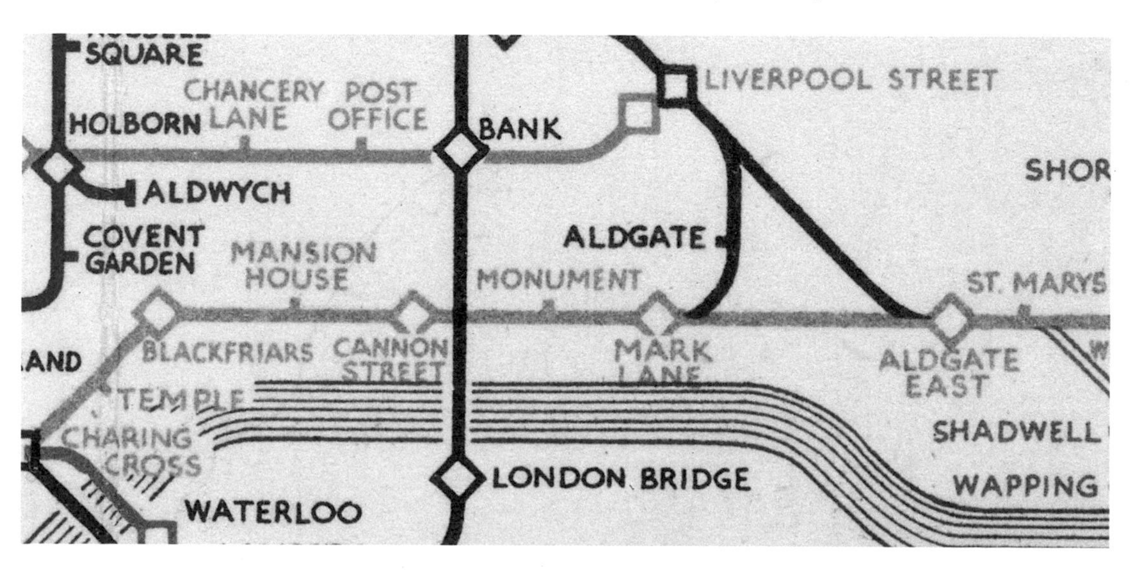

Chapter Four

*W*inning the *D*ay

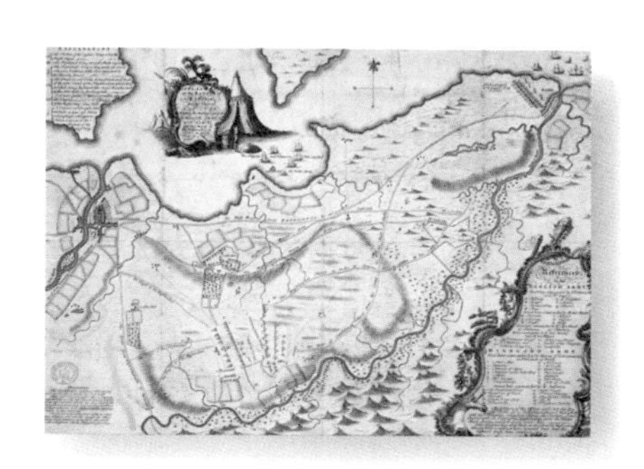

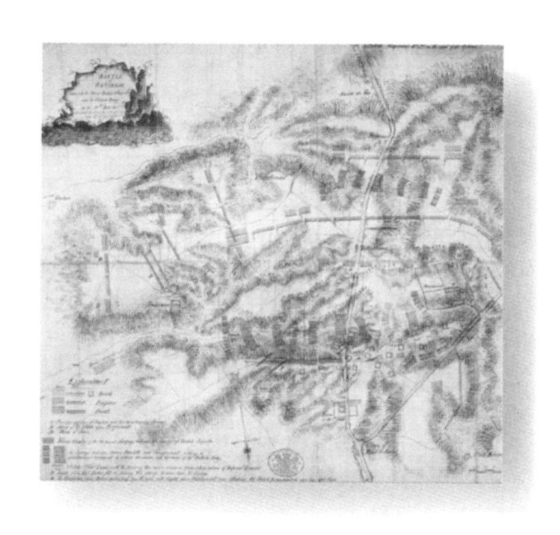

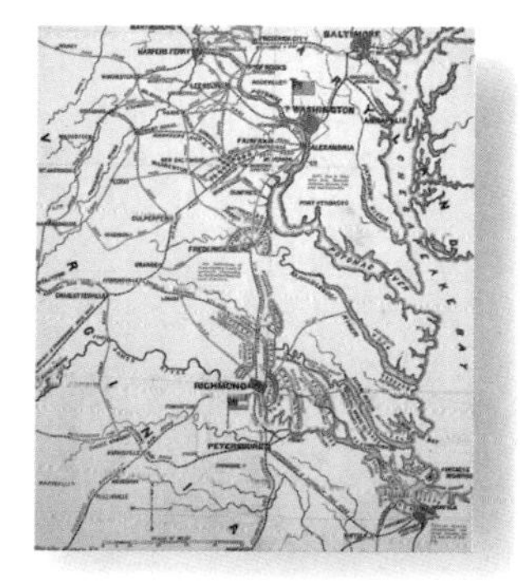

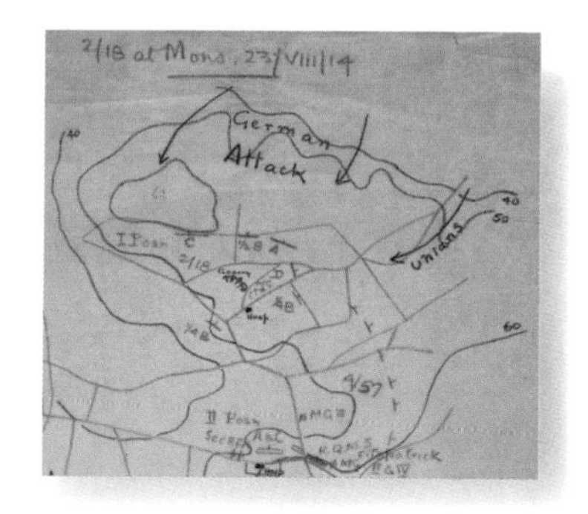
2/1B at Mons, 23/VIII/14
German Attack
Uhlans

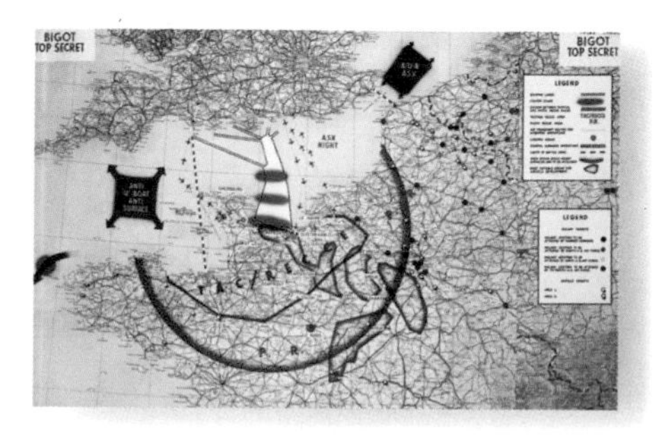
BIGOT
TOP SECRET
BIGOT
TOP SECRET

The link between maps and the military has always been strong. The roots of many countries' national mapping agencies can be traced back to military origins, such as the Ordnance Survey in Great Britain, which dates back to the French Revolutionary War. The military link is not surprising: the need to move large amounts of personnel and equipment over considerable distances requires information about mileage, routes and difficulties presented by the terrain. Add to that the menacing presence of an opposing force and the need for such information often becomes crucial to survival.

Alongside the movement of forces there was the issue of where to place them for a forthcoming battle. Without maps generals were limited to what they could actually see or what others reported to them. As will be seen in the discussion on the Battle of Waterloo (pp. 128–33), the Duke of Wellington's access to a detailed map prior to the battle gave him a definite advantage which he exploited to the full as he positioned his troops.

Once the conflict had begun one might suppose that the map had played its part but this was not so. The value of making a record of the battle itself, including the timing and movement of different types of troops, provided the data for the analysis that increasingly took place after the event itself. As professional armies replaced the armed militia of an earlier era a more scientific approach to warfare developed with the requirement for post-battle analysis. The collection and recording of information as the battle raged was a particularly dangerous role and one that has hardly been accorded sufficient status. It is appropriate that the role of Captain Thornton of the Royal Engineers in the creation of the Waterloo map is noted on it.

However, creating such a map would not necessarily be soley for military analysis. Unlike some of the famous Caesars and emperors from antiquity modern democracies require a degree of

60 pieces of Cannon, belonging to the Imperial Guard and several Carriages, Baggage &c. belonging to Bonaparte in Genappe.

I propose to move this morning upon Nivelles & not to discontinue my operations.

Your Lordship will observe that such a desperate action could not be fought and such advantages could not be gained without great loss, and I am sorry to add that ours has been immense. In Lieut. Genl. Sir Thomas Picton His Majesty has sustained the Loss of an Officer who has frequently distinguished himself in his Service, and he fell gloriously leading his Division to a charge with Bayonets by which one of the most serious attacks made by the Enemy on our Position was defeated. The Earl of Uxbridge after having successfully got through this arduous day received a

Correspondence from the Duke of Wellington to the Earl of Bathurst gives his account of the Battle of Waterloo in 1815.

accountability from their military leaders. Politicians expect to be provided with information about how and when the military ventures they have sanctioned actually took place, while the general public want to be able to 'follow the war', and maps have become increasingly important as the main means of facilitating this. The front page of *The London American* newspaper of 1861 publishing information about the initial battle of the Civil War (pp. 134–9) is an early example of this development.

The maps chosen for inclusion in this chapter each provide examples of the different but overlapping roles of the map in military operations. They enable us to develop a better understanding of the battles, large and small, that they refer to.

Whether Charles Stuart's access to a better map (pp. 122–7) would have had any impact at Culloden when he and his clansmen attempted to reclaim the Scottish throne from the English is debatable but unlikely. His failings as Commander-in-Chief on that fateful day in April 1746 were more comprehensive. It is reported that one of his clan chiefs pointed out that the gently sloping and relatively even ground of Drumossie Moor suited the English cavalry better than charging highlanders but he dismissed the significance of this. A map showing this same information is unlikely to have convinced him otherwise.

The map showing the battle for the small hillock just south of the Belgian town of Mons in 1914 (pp. 140–45) is at the other end of the scale from a Waterloo, Bull Run or Culloden but if anything this makes it more poignant. The story of this map recounts the disarray and confusion that developed over a 24-hour period as the remnants of one British regiment gladly joined up with the remains of others in the attempt to simply stay alive. One can imagine this very map being consulted by the senior officer in charge of the survivors of the Royal Irish Regiment as they found their escape route suddenly blocked by German troops in the late afternoon. Without the information it showed he may well have not made the decision to take the route that did eventually take most, but not all, of them to safety. At least for that day.

While the hand-scribbled notes and the arrows on the map of Bois la Haut convey the urgency and dynamics of the day's events so the D-Day map (pp. 146–51) does quite the opposite. Drawn after the events of 6 June 1944 to create a lasting record of how one branch of the military played its part in securing the success of Operation Overlord it is clear and precise. As such it is informative but also deceptive. When one combines the information it presents with the personal accounts of those involved it becomes clear that many gliders and planes carrying the parachutists did not arrive at their destinations by flying in the straight lines shown here or even land in the marked area. The tactical 'recce' (reconnaissance) area marked gives the impression that the Allies knew all that was necessary about what it contained. Accounts from the ground forces show different as many surprises were encountered as they fought their way inland.

None of this invalidates the map itself or the credibility of the map's creator. What it serves to do is remind the reader once again that maps on their own can never give a complete picture and need to be used in conjunction with other pieces of evidence.

56
EXPLANATION.
A.A. The Position of the English Army when the Battle began.
B.B. The Highland Army when the Battle began
C.C C.C The Highland Army as they made the Attack only some breakings in the Line occasion'd by the Marshy Ground.
D.D. The Dragoons of ye Dukes Army form'd on ye Rear of the right Flank of the Highlanders where they marched during the Attack thro' some Breaches made in the Stone Walls by the Campbells.
E.E. Lord Lewis Gordons & Ogilvie's form'd to oppose the Dragoons which favour'd ye Retreat of the Right Wing when broke
F. A Cannon not brought into the Field in time of Action which play'd on Kingstons Horse & favour'd the Retreat of the Left Wing
A PLAN of the BATTLE of Culloden and the Adjacent COUNTRY Shewing the Incampment of the ENGLISH ARMY at NAIRN, and the March of the Highlanders in Order to Attack Them by Night.
W
E
S
The Shipping and Transports that supplyed and attended the Dukes Army
INVERNESS
The High Road from INVERNESS to NAIRN
Culraick
Culraick
The Highlanders Return from
The Highlanders Return from
Culloden
The Highlanders March in Order to Surprise the
to the Field of Battle
The Duke of Cumberlands March from his Camp
Ness River
Road from Inverness to Fort Augustus
The Flight of the Highlanders
The Flight of the Highlanders
The Flight of the Highlanders
The Flight of the Highlanders
Castle Hill
The Road from Inverness to Ruthven
PUBLIC RECORD OFFICE
Observations
The Highlanders had been without pay, and scarce of provisions for some weeks.
They were obliged to fight after a fatiguing march, without any refreshment: having had no sleep and but little food the two days and nights immediately preceeding; and wanting numbers of their men, who were dispers'd in the adjacent villages on these accounts.
It was but a small number of them, (that were present,) did actually engage, the others being intimidated on seeing those who made such a desperate attack, obliged to give way.
112

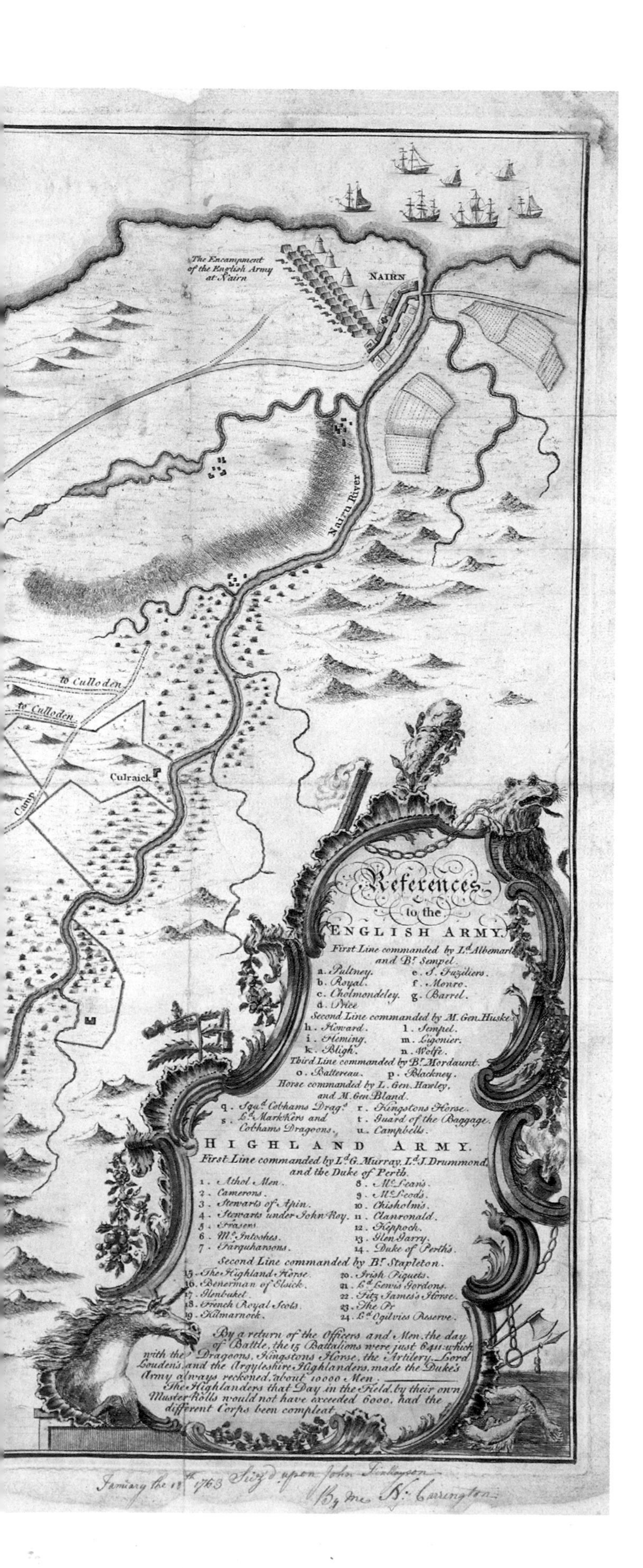

The Last Battle

The Battle of Culloden

By an unknown soldier, 1745

The Last Battle

The Battle of Culloden

This precise and intricate map is central to the story of a nation. The battle detailed here saw Scotland's claims to being an independent nation under its own monarch smashed in a bloody and one-sided contest. It was to take nearly 250 years before a degree of political independence would be re-established through political rather than military means.

Culloden was the last battle fought on the British mainland and it was not against a foreign invader but was rather the last action in a long and sporadic civil war. The final stages of this began in 1745 when Charles Edward Stuart landed at Arisaig to reclaim the crown he believed was rightly his. He had seven men with him when he landed but he was confident that the ancient clan system, of which he was the ultimate head, would provide the army he needed. His confidence proved justified and most, but not all, of the chiefs brought their clansmen to his camp. Throughout 1745 Charles was spectacularly successful and marched south as far as Derby before turning back to Scotland. The English assembled a well-equipped and disciplined army who marched into the Highlands of Scotland in pursuit.

The natural 'home advantage' of the Highland army was negated by incompetent leadership. Charles had some excellent generals but for whatever reason – probably a combination of vanity and overconfidence – he decided to take full command of this encounter. He determined to mount a surprise attack on the English army while it was encamped in Nairn on the north-eastern coast. However, he underestimated how long it would take his army to march there and by the time they approached the town it was too close to dawn – so they marched back to Culloden. They had been marching most of the night. Also, despite the fact that there were supplies in Inverness, delivery had not been organized and most of his army had had nothing more than a biscuit to eat in the 48 hours prior to the battle. He was also awaiting reinforcements. The combination of a tired, hungry and under-strength army might have prompted caution, but not in Charles. He lined his men up on Drumossie Moor just beyond Culloden

House, as shown on the map, on the morning of 16 April 1746. Despite their condition some 6,000 Highlanders dressed in their plaids (a long piece of woven wool wrapped around their thighs as a skirt) and armed with the broadsword and targe (shield) would have made a fearsome sight.

The Duke of Cumberland's army, dressed in traditional red uniforms and armed with musket and bayonet, marched from Nairn to face them. On the day they were in far better physical shape, for it had been Cumberland's birthday the previous day and he had ordered extra rations and drink to celebrate. Cumberland was not a particularly great military leader; but at Culloden he didn't need to be.

One of the enduring mysteries of the first stage of the battle is why the order to charge was not given to the clansmen. The superior English artillery, clearly shown on the map, fired round after round into the assembled ranks as they stood waiting the order. Hundreds were killed or maimed without having used their weapons. Finally, impatience and frustration took over and a charge began and spread down the line. The English cannons exchanged their cannon balls for grapeshot and simply fired straight ahead. Alongside this, the discipline of the English musket men firing in volleys, moving aside and back to let the next row fire while reloading took a terrible toll. While some clansmen did make the English front line, the numbers were insufficient to break through. Individually they had no chance.

Artistic impression of the Battle of Culloden, looking across Drumossie Moor as the Scottish clansmen begin to charge the English line.

The battle lasted just over an hour. In this short time an estimated 2,000 clansmen were killed compared to only 50 in the English army. No mercy was shown to the defeated, and the wounded were executed where they lay while those who escaped were hunted down and killed for being 'rebels and traitors'. This gave rise to the English General being nicknamed 'Butcher' Cumberland, a name he is still known by in Scotland.

Not all battles prove to be significant. This one, however, was extremely significant for Scotland and rarely can the longer-term consequences of a single battle have been greater. The ancient clan system was systematically demolished by the English, the carrying of a sword became punishable by death, while the wearing of the plaid and clan tartan, and even the playing of the bagpipes, became punishable offences. Huge areas of Scotland were emptied of their inhabitants in what we would now recognize as ethnic cleansing. The origin of the many Scottish communities found across the world can be dated to this period as the diaspora sought to rebuild their lives in other places.

But, as is often the case, heroic defeat gives rise to myth and legend and Charles Edward Stuart is better known today as the romantic figure of Bonnie Prince Charlie, immortalized in the lines of the Skye Boat Song:

Speed bonnie boat like a bird on the wing,
Onward the sailors cry,
Carry the lad that's born to be king,
Over the sea to Skye.

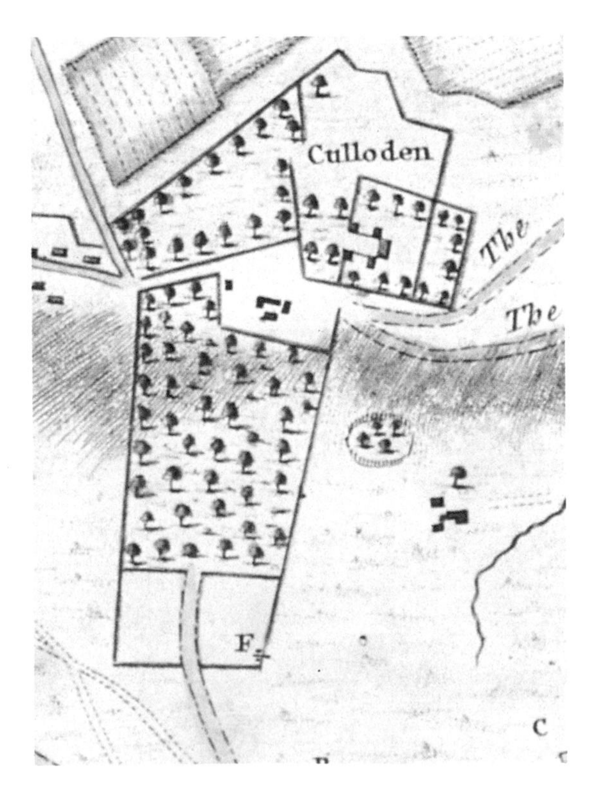

Culloden House

The battle itself was fought on Drumossie Moor, but it was the stately home that lay over a mile to the north that gave its name to posterity. As can be seen on the map, it lay just outside Inverness off the coastal road to Nairn.

In mid-April 1746 Charles took over the house as his headquarters. The ill-fated 'surprise' attack on the English army was planned here and, as the map shows, it was to this place that the exhausted and hungry clansmen returned when it was aborted. They would have found what shelter they could in the woods before heading out of the southern gate to assemble on the moor.

Culloden House is now a luxurious hotel set in beautiful grounds. Although rebuilt in what might be considered an aptly named 'Georgian' style some time later, many of the features of the house that witnessed the epic Battle of Culloden on 16 April 1746 remain. The vaulted cellars are likely to have witnessed some horrific scenes as Cumberland's troops came across the wounded clansmen seeking refuge there.

The Clans Assemble

As they marched on to Drumossie Moor on the morning of 16 April the Highland army lined up clan by clan. The 'References' in the bottom right-hand corner of the main map carefully identifies these one by one. The order in which they formed up was not arbitrary. It wasn't 'chance' that found the Athol Men (No. 1 on the map) in the position they were and this placement had been the cause of serious dispute in Charles's army. The right flank was a place of honour and the Clan MacDonald had claimed this position as its own ever since the famous victory over the English at the Battle of Bannockburn in 1314. However, Charles's key supporter, Lord George Murray, had persuaded him to allow his men this prestigious position. It proved to be a fatal choice. The holders of this right flank were the closest to the withering musket fire that rained on them from behind the enclosure wall once the charge began. Very few of them actually made it to the English line. The MacDonalds suffered a similar fate at the other end of the line as they were first decimated by grapeshot from the English cannon, then cut down by volleys of musket fire as they charged, before being pursued by the cavalry of Kingstons Horse as they retreated. Those clans in between the Athol Men and the MacDonalds fared little better.

Nevertheless, the clan names listed here remain evocative and even familiar to us today. Each is proud to recall its presence at Culloden even after 250 years.

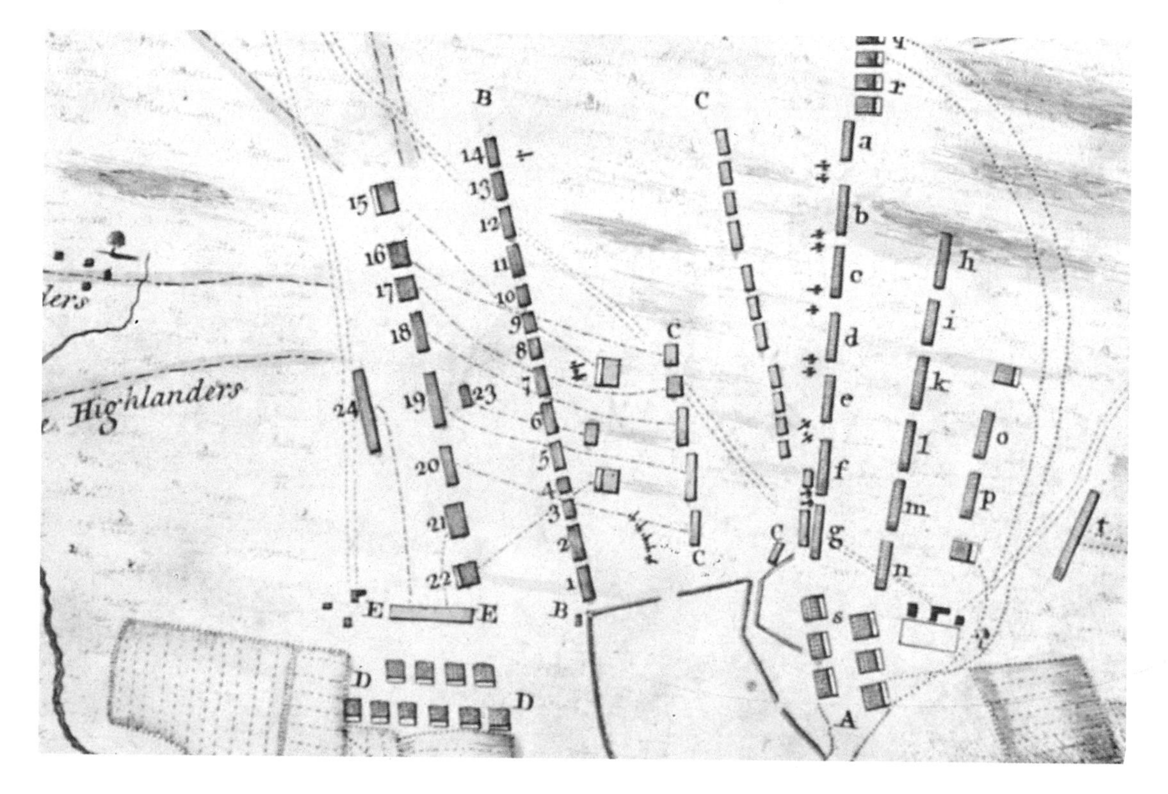

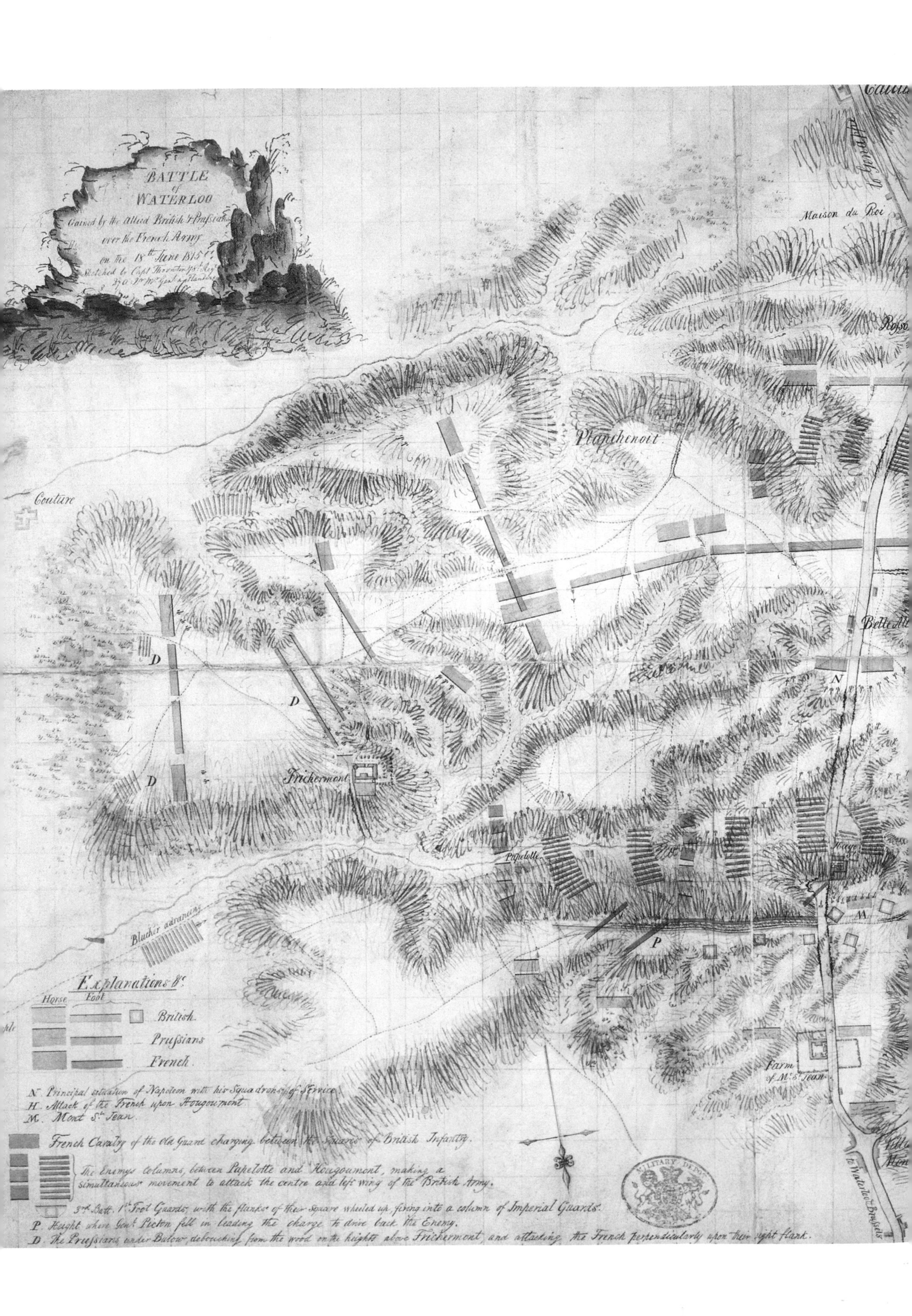
BATTLE
of
WATERLOO
Gained by the allied British & Prussians
over the French Army
on the 18th June 1815
Maison du Roi
Planchenoit
Couture
Frichermont
Papelotte
Belle Al
N
D
D
D
P
M
Farm
of Mt. St. Jean
Blucher advancing
Explanations &c.
Horse
Foot
British
Prussians
French
N. Principal situation of Napoleon with his Squadrons of Service
H. Attack of the French upon Hougoumont
M. Mont St. Jean
French Cavalry of the Old Guard charging between the Squares of British Infantry
The Enemys columns, between Papelotte and Hougoumont, making a simultaneous movement to attack the centre and left wing of the British Army.
3rd Batt. 1st Foot Guards, with the flanks of their square wheeled up, firing into a column of Imperial Guards.
P. Height where Genl. Picton fell in leading the charge to drive back the Enemy.
D. The Prussians under Bulow, debouching from the wood on the heights above Frichermont, and attacking the French perpendicularly upon their right flank.
MILITARY DEPOT
to Waterloo & Brussels

A Damned Near-Run Thing

The Battle of Waterloo

By Captain Thornton, *c.*1815

A Damned Near-Run Thing

The Battle of Waterloo

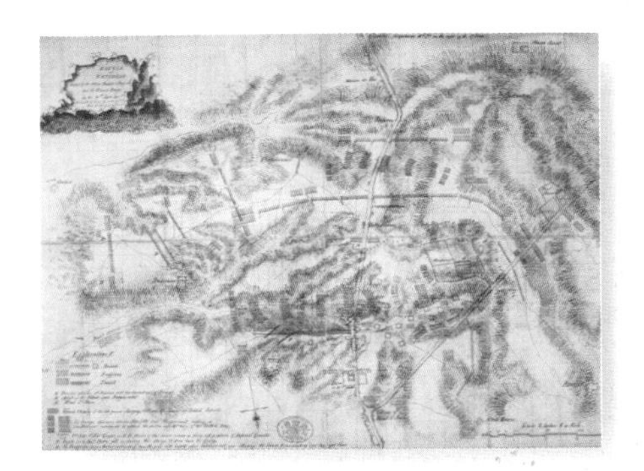

Wellington, Napoleon and Waterloo. The use of any one of these names in a conversation will almost inevitably lead to a reference to one of the others. This is not surprising as the destiny of all three were inextricably linked. Before the battle that took place on 18 June 1815 very few people would have known of the existence of the small Belgian town of Waterloo. And yet since that fateful day its name has become so familiar through its use as the name of a major bridge over the River Thames, as the name of one of London's principal railway stations and even for a song that won the European Song Contest! The phrase 'met his Waterloo' is an expression that continues to be used when a final reckoning has taken place for an individual or institution. The fate of Napoleon, once the most powerful man in the world, was certainly sealed here. His subsequent exile to the island of St Helena was to prove permanent and he died there in 1821. Waterloo was also the Duke of Wellington's last battle and, after acting as Commander-in-Chief of the occupation of France, he returned to England as a national hero in 1818. Ten years later he was to become Prime Minister, although perhaps not with the same degree of success. However, his place in history was already secure.

A close examination of the map shows that the military features have been drawn on to a pre-existing map. The original had been surveyed and drawn only the previous year as part of a broader survey ordered by Wellington. He could have had no way of knowing how significant it would become. It was on a copy of this map that the British officer, Captain Thornton of the Royal Engineers, made this record of what was to prove one of the most significant battles of the nineteenth century. Thornton's carefully drawn lines and blocks, each painstakingly colour coded to show the Allied, French and Prussian armies and their infantry, cavalry and artillery, are informative and deceptive at the same time. The neatness of the squares and columns gives a false impression of order. Once the battle commenced, confusion often verging on chaos would frequently take over and successful communication was difficult to achieve. As Wellington or Napoleon sent yet another rider off with fresh orders neither could be sure they would arrive at their destination. Even when they did neither could be sure their instructions would be interpreted accurately or that the bugles and drums that conveyed the message to the troops would be heard above the noise of the battle. Finally, there was always the possibility that they would not be obeyed.

Portrait of Arthur Wellesley, 1st Duke of Wellington, by Sir Thomas Lawrence, 1814.

At Waterloo Wellington had used the information from his earlier survey to place his main army just over the brow of a ridge. As the map shows he had chosen the location so that he had a village and a large house at either side to protect his flanks and the stone buildings of La Hay Sainte in the centre of his line.

Thornton's map conveys very successfully that it was the French army who took the offensive. The numerous sets of parallel blue lines, often bending to the right or left are French brigades on the attack. Although quite faint, it is just possible to make out the dotted line he has drawn behind each of these units that shows the positions they have advanced from. Napoleon had no alternative but to go on to the attack for he knew he had to secure victory before the allied armies could join together. The Prussian force under Field Marshall Blücher was only 8 miles (13 kilometres) away. However, when, as Thornton noted, they 'debouched from the wood on the heights' (annotation in bottom left corner) to attack the French flank it proved decisive.

Approximately 130,000 men faced each other at the beginning of the day but 15,000 English and Dutch, 7,000 Prussians and nearly 30,000 French didn't live to see it end. The scale measure (bottom right) shows this is a large-scale map and this is a little deceptive in creating a somewhat spacious feel to the battlefield. There are a number of personal accounts from survivors and all comment on the feeling of being packed together and unable to move. The second French cavalry charge of the day consisted of nearly 5,000 horsemen over a line of barely 654 yards (600 metres) and one who survived recorded that he felt his horse being carried along without its hooves touching the ground!

The Battle of Waterloo brought Napoleon's imperial aspirations to an end. But as Wellington is recorded as saying after the battle it was 'a damned near-run thing'.

Within Sight of Each Other

Besides showing the different elements of the two armies, Thornton has also marked on the positions of the two opposing generals. 'N' locates Napoleon and 'W' Wellington. Both commanders would have sought a position which gave them an elevated vantage point over the battleground but which also enabled them to communicate with all units quickly. As the map also shows, this meant they located themselves centrally at Waterloo. While this had definite advantages it also made them obvious targets for their opponent's artillery. Using the scale provided it is possible to establish that Napoleon and Wellington were less than 1 mile (1.6 kilometres) apart throughout the battle. It is therefore very likely that they could actually see each other through their telescopes!

Hougoument

The small Chateau of Hougoument proved to be a critical location in the Battle of Waterloo. Wellington had positioned his army so that the chateau and its walled and wooded grounds formed the right flank of his position. Colonel James MacDonnell and units of the Coldstream Guards were given the responsibility for its defence.

Napoleon was clearly in agreement with Wellington's assessment that this was a crucial position and he directed his initial attack here. The French under Prince Jerome quickly breached the walled gardens but despite sustained attacks right through the day the house itself was not taken. By late afternoon the surrounding area was covered with the dead, with estimates of nearly 10,000 soldiers losing their lives in this sector of the battlefield. The stubborn defence of Hougoumont tied up large numbers of Napoleon's best troops and his failure to unhinge the allied army on this flank was a decisive feature of the day. This extract captures the intensity of the fighting around the Chateau with Thornton drawing the blue of the French units touching the red representing the British forces.

The significance of Hougoumont to the day's events is reinforced by a story from some years later. An English churchman donated the large sum of £500 to be 'given to the bravest man at Waterloo' and Wellington was asked to adjudicate. He chose Colonel MacDonnell for his defence of the chateau. However, MacDonnell insisted on giving half on this sum to a Sergeant Graham who had fought alongside him to defend the main door when it seemed the French were certain to break through.

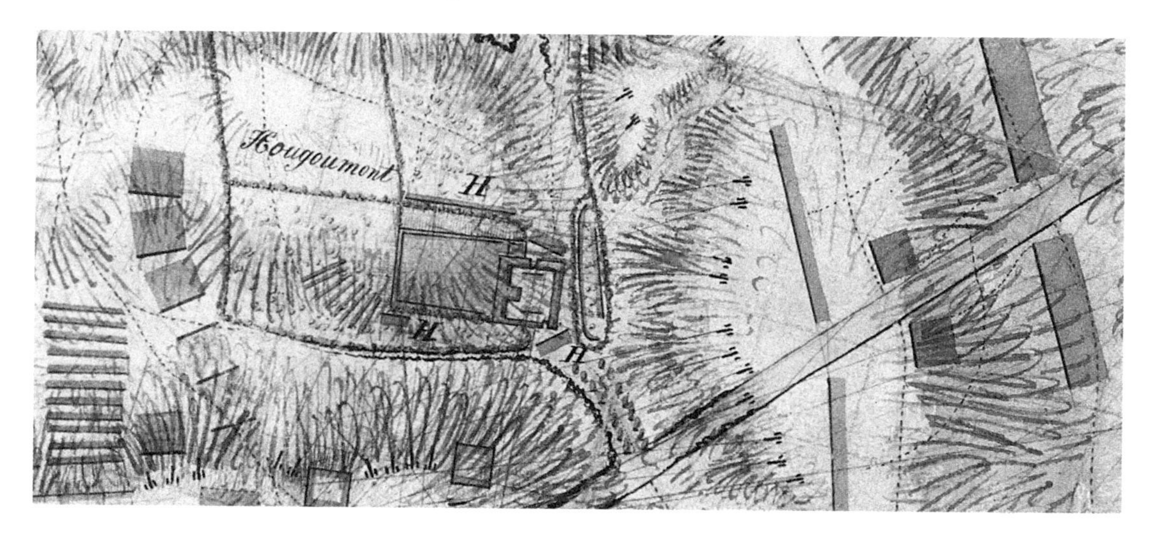

Late But Decisive Arrivals

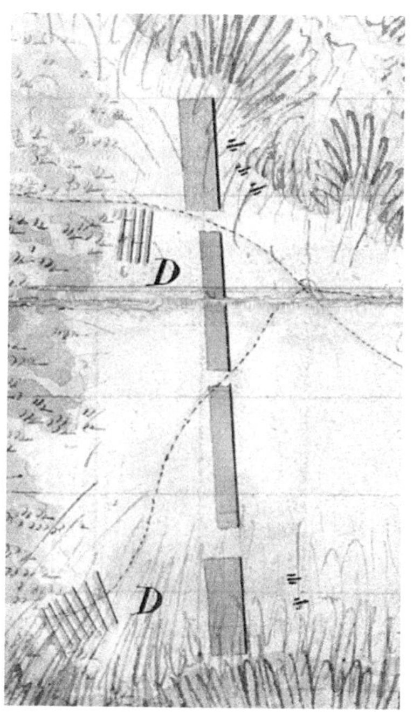

The significance of the arrival of the Prussian army on the battlefield of Waterloo continues to be debated. Only two days earlier Blücher's army had been badly beaten by Napoleon at the Battle of Ligny and the Emperor was confident that the Prussian army would play no part as he confronted Wellington. However, it was the weather rather than tactics that almost decided this. A fierce thunderstorm during the night had turned the roads into quagmires and it was late afternoon before the first Prussian reached the battlefield. It was apparently this same thunderstorm that had prompted Napoleon to delay his first attack until one o'clock in the afternoon 'so that the ground could dry'. If the battle had commenced earlier it would almost certainly have been over by the time the Prussians arrived.

As this extract clearly shows the Prussians would have been very visible as they came out of the woods high on the hill to the east. With the French and Allied armies having fought themselves almost to a standstill the psychological impact of the new arrivals was immense. Napoleon was keenly aware of this and actually put a rumour into circulation that these were in fact French reinforcements. However, the direction of their fire quickly established the reality of the situation.

The London American.

MAP OF THE SEAT OF WAR,

POSITIONS OF THE REBEL FORCES, BATTERIES, ENTRENCHMENTS AND ENCAMPMENTS IN VIRGINIA—THE FORTIFICATIONS FOR THE PROTECTION OF RICHMOND.

THE "LONDON AMERICAN," AN INTERNATIONAL NEWSPAPER, PUBLISHED EVERY WEDNESDAY MORNING, PRICE 3d. OFFICE, 9, EXETER CHANGE, & ALL NEWSMEN.

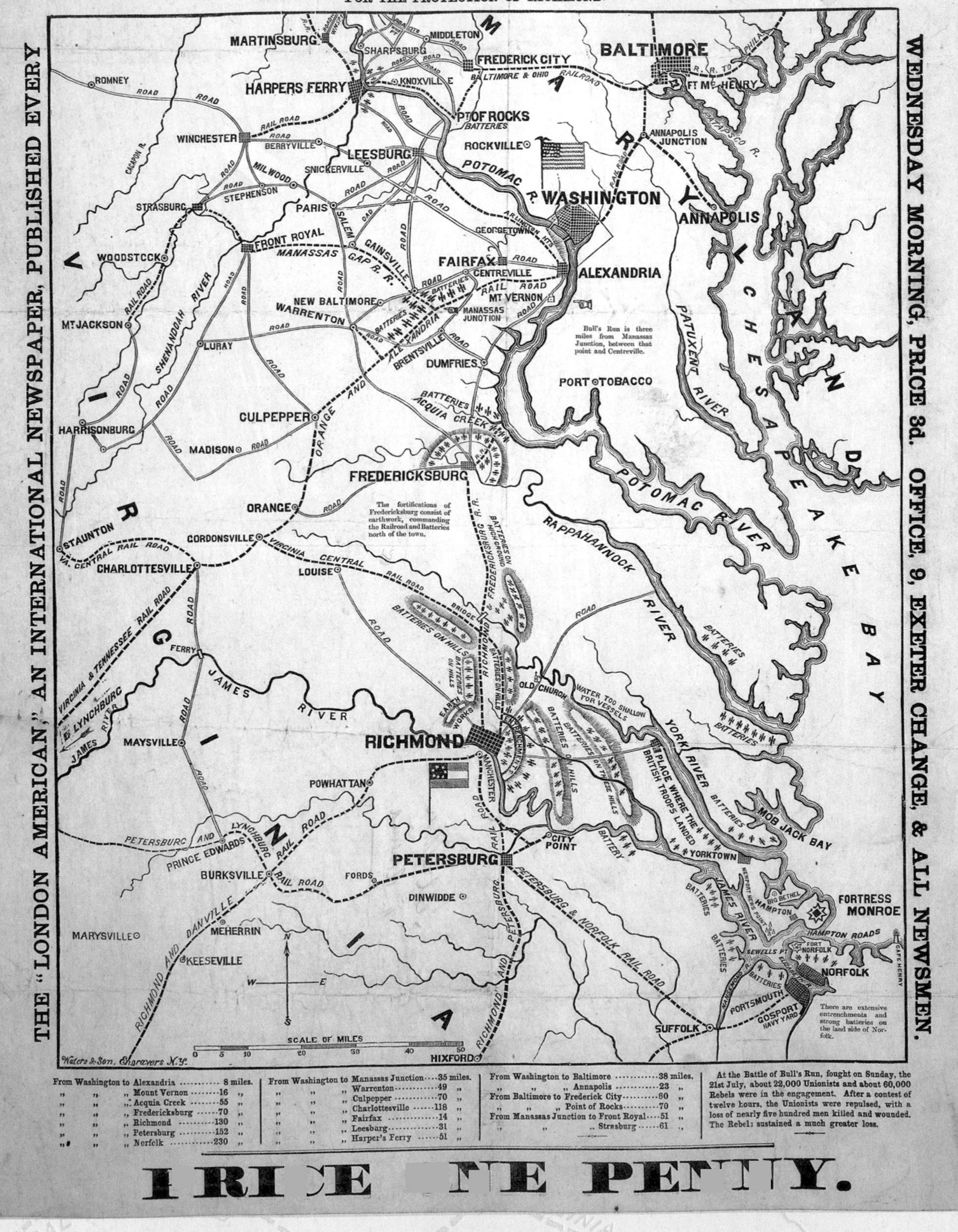

From Washington to		Miles
Alexandria		8 miles.
Mount Vernon		16
Acquia Creek		55
Fredericksburg		70
Richmond		130
Petersburg		152
Norfolk		230
Manassas Junction		35 miles.
Warrenton		49
Culpepper		70
Charlottesville		118
Fairfax		14
Leesburg		31
Harper's Ferry		51
Baltimore		38 miles.
Annapolis		23

From Baltimore to	Miles
Frederick City	60
Point of Rocks	70

From Manassas Junction to	Miles
Front Royal	51
Strasburg	61

At the Battle of Bull's Run, fought on Sunday, the 21st July, about 22,000 Unionists and about 60,000 Rebels were in the engagement. After a contest of twelve hours, the Unionists were repulsed, with a loss of nearly five hundred men killed and wounded. The Rebels sustained a much greater loss.

PRICE ONE PENNY.

In the News

The American Civil War

Published in *The London American,* 1861

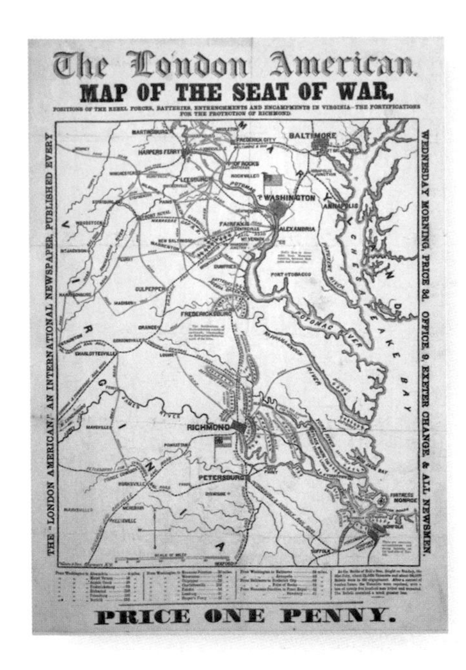

In the News
The American Civil War

In the modern technological age details about any conflict anywhere in the world are readily available. We have become used to seeing reporters in protective clothing crouching down on the edge of the battlefield as they provide 'updates'; familiar with generals giving interviews from their headquarters using satellite maps to illustrate their points; and accustomed to presidents and prime ministers seeking to explain the political implications of the military events in televised broadcasts. This is a very new situation. However, the need to know what was happening during wartime and to base decisions on the information available has existed since ancient times and that was the purpose of this map.

It dates from 1861, the year the American Civil War began and, as the layout confirms, it formed the front page of a newspaper. *The London American* had a short life span being first printed in May 1860 and ceasing to be published in March 1863. During this time it played an important role in informing the British public and politicians of events over 3,000 miles (4,830 kilometres) away and at least two weeks old. However, like all newspapers it also sought to influence the reader through the way it presented the news and this is a good example of how subtly this can be undertaken. The use of 'rebel' in the subtitle is indicative of *The London American*'s pro-Unionist position. It supported the newly elected President, Abraham Lincoln, in his determination to keep the American states united and opposed the secession of the 11 southern states, the Confederacy. It was keen to persuade London to view the conflict in the same way.

In 1861 there was probably a greater chance of Britain siding with the Confederacy than with the Unionist side. The political power of the aristocracy was still considerable and there was a clear feeling of affinity with the Southern landowners and a real suspicion of the assertive populist democracy of the North. However, popular feeling would have made it very difficult for the British Government to have openly supported the side that was fighting to preserve slavery.

Nevertheless, for the first two years of the conflict it looked as if the Confederacy would actually win. In late 1862 William E. Gladstone, the British Chancellor of the Exchequer, is on record as saying, 'We may anticipate with certainty the success of the Southern States so far as regards their separation from the North'. Those, including *The London American*, who sought to advance the Unionist cause clearly had an uphill task. At the Battle of Bull Run as the newspaper refers to it, or the Battle of First Manassas as it has become known, the Union army was badly defeated and retreated back into Washington in disarray. The carnage and bloodshed of this first encounter ended the notion that this would be a dispute that could be ended quickly and prompted Abe Lincoln's comment, 'It's damn bad'.

The offer of a bounty was common practice in encouraging men to enlist, as this poster shows.

What the map shows is the manner in which the Confederacy had fortified its position, particularly around its capital city of Richmond. It reflects the defensive strategy adopted by the Confederacy and the onus this placed on the Union to adopt an offensive one in its attempt to prevent secession. As can be seen, the direct route from the north was barred by the fortified town of Fredericksburg and the numerous gun batteries on the hills that formed a corridor leading south to Richmond itself. However, as the map also reveals, the many rivers and inlets leading off from Chesapeake Bay presented the real threat of an army being landed to the east or south of the town. The numerous gun emplacements drawn on the banks of the James and York rivers clearly reflect the awareness of this threat. However, they proved unable to prevent the seaborne Union army landing the following year and marching on the Confederate capital from the south-east. Only the inspired general-ship of Robert E. Lee, taking control of the Southern forces for the first time, prevented Richmond falling.

The map was drawn and published barely four months into a war that was to last for four years and yet the title given to it by an unknown subeditor proved prophetic. This area was indeed to prove the 'seat' of the conflict with over 3,000 military engagements taking place in the states of Virginia, West Virginia and Maryland, although very few of these were large-scale battles like the one at Bull Run. What no one who looked at it in 1861 could have realized is that it also showed the location where the conflict would end a few, bloody, years later.

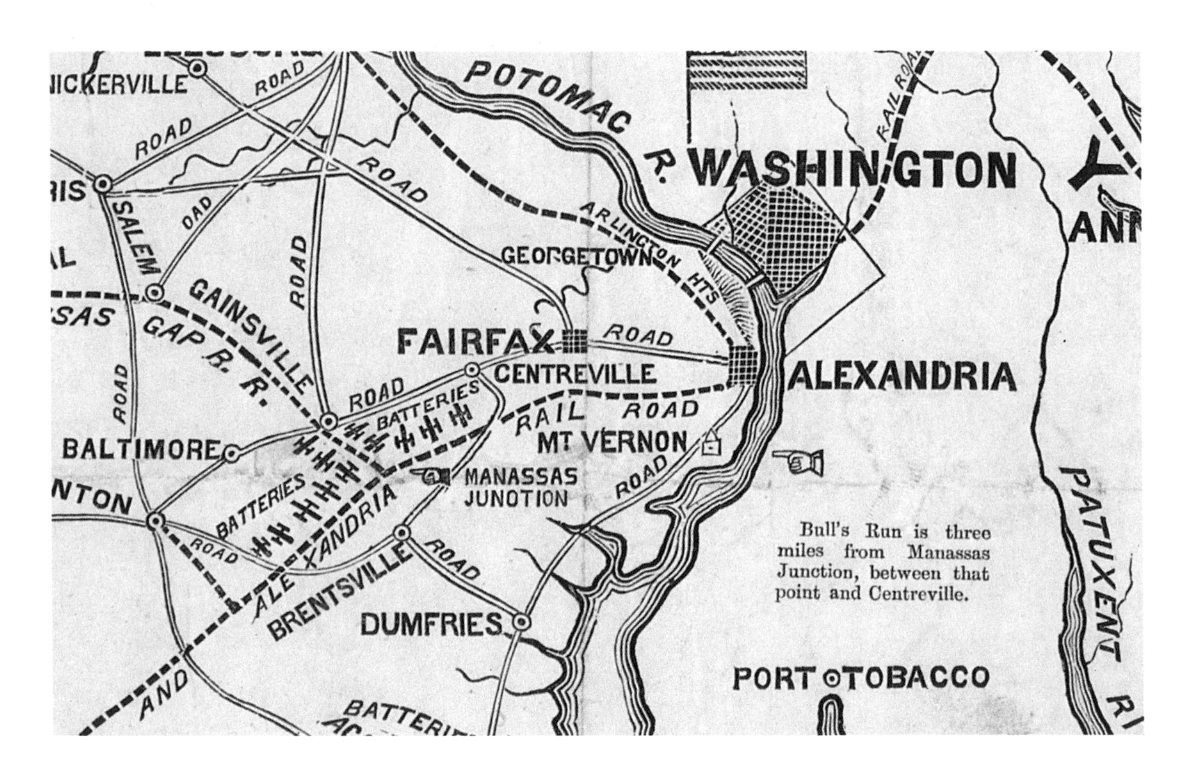

The Battle of Bull Run

Whether referred to as The Battle of Bull Run or The Battle of First Manassas, this was the first full pitched battle of the war. As the extract above reveals it was a highly significant location, because whoever controlled this area controlled the rail gateway to the South. In the era before motorized transport the railway was the fastest means of transporting large quantities of men, equipment and supplies over any distance. When war became imminent, Manassas Junction, only 35 miles (56 kilometres) from Washington immediately assumed strategic importance. The cannons shown on the map are only representative but certainly convey the very real deterrent to any attempt by the Union to move its forces south by this route. *The London American*'s political

At the Battle of Bull's Run, fought on Sunday, the 21st July, about 22,000 Unionists and about 60,000 Rebels were in the engagement. After a contest of twelve hours, the Unionists were repulsed, with a loss of nearly five hundred men killed and wounded. The Rebels sustained a much greater loss.

sympathies are conveyed through the small block of text in the bottom right-hand corner of this front page (see above). Here it reports the result of this major engagement in a very low-key manner, thus attempting to underplay its significance. Its deliberate misrepresentation of the respective sizes and losses of the two armies further contributes to this effect. In fact both armies were close to 30,000 men with the Union casualties being in the region of 3,000 to the Confederacy's 2,000.

Such was the strategic significance of this location that there was to be a second and even fiercer battle in August 1862 with the Union side again being defeated, this time with even greater losses.

The Seat of War

This newspaper front page was produced at the beginning of the war and the map details the events of these early months. What could not be known at the time is that this very map also contained the location where the war would end. If one traces the route of the Petersburg and Lynchburg Rail Road as it runs west and finds the point where it crosses a small unnamed river one has identified the site of the tiny village of Appotomax Courthouse. Here the exhausted and outnumbered Confederacy army of General Lee were finally caught by the encircling Union forces of General Ulysses Grant. Grant accepted Lee's surrender at Appotomax Courthouse on 9 April 1865 and one of the bloodiest conflicts of the century came to a halt. Overnight this remote hamlet became one of the most famous locations in American history. The original title given to this map in 1861 had proved to be an appropriate one, for the area it featured did indeed prove to be 'The Seat of the War'.

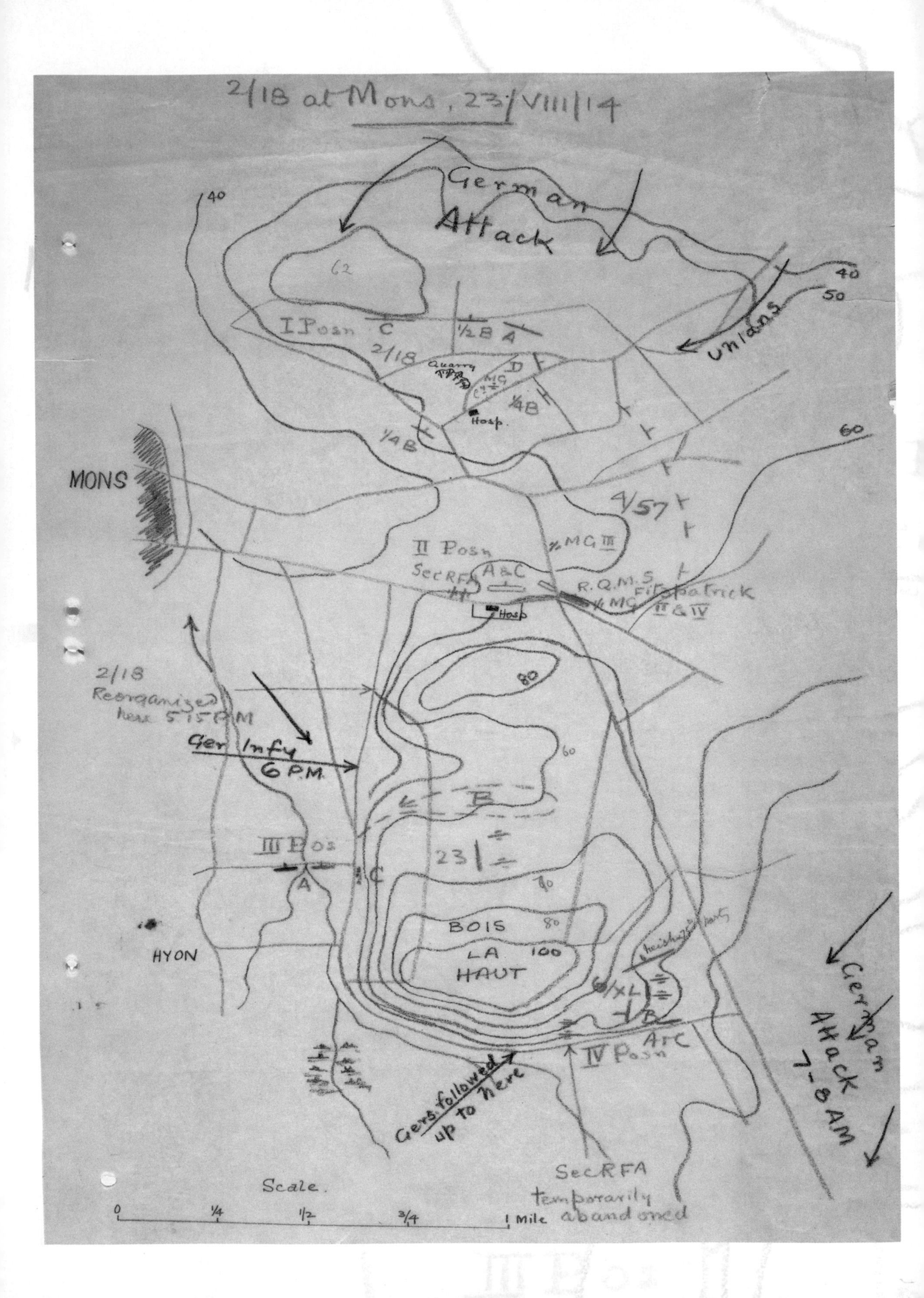

2/18 at Mons, 23/VIII/14
German Attack
40
62
40
50
Uhlans
I Posn
C
1/2 B
A
2/18
Quarry
D
MG
1/4 B
Hosp.
1/4 B
60
MONS
4/57
II Posn
MG III
Sec RFA
A & C
R.Q.M.S. Fitzpatrick
MG
II & IV
Hosp
80
2/18
Reorganized here 5.15 P.M
Ger Infy
6 P.M.
60
B
III Pos
23
A
C
70
BOIS
80
LA
100
HAUT
HYON
XL
B
A+C
IV Posn
Gers followed up to here
German Attack 7-8 AM
Scale.
0
1/4
1/2
3/4
1 Mile
Sec RFA temporarily abandoned

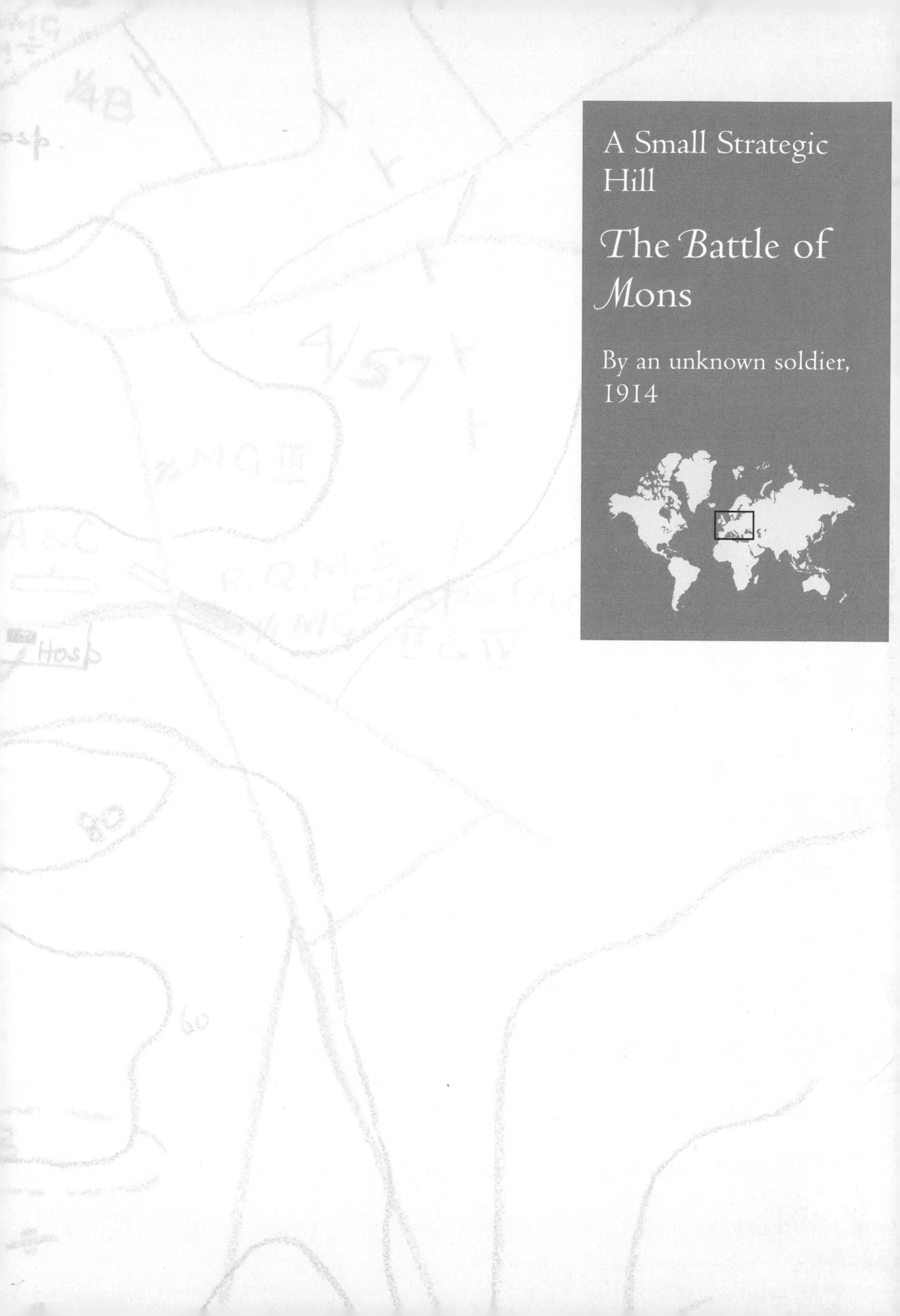
A Small Strategic Hill
The Battle of Mons
By an unknown soldier, 1914

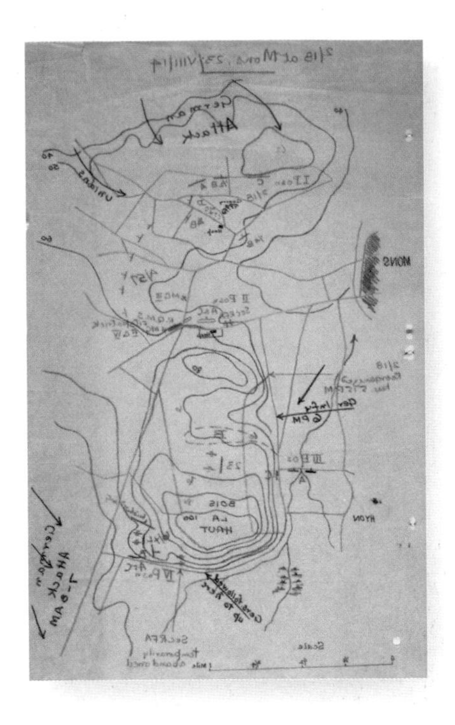

A Small Strategic Hill
The Battle of Mons

Thousands of maps were drawn up by all nations involved in the First World War. The National Archives alone holds over 6,000 relating just to the Western Front. The scale of these maps varies from those that are continent wide providing the 'big picture', to those like the one here which details what happened in a small geographical area on a specific day. The map takes the viewer closer to the actual experiences of the soldiers who were there by detailing some of the small-scale events that made up the battle. Indeed, it is small-scale actions such as these by individual soldiers that have often determined the course of battles throughout history.

As the hand-written date across the top indicates it records the events that took place near Mons on 23 August 1914. The date is very significant. It was the day on which the first full encounter between the British and German armies took place. The conflict that began in late August would continue for a further four years and neither side could possibly have foreseen that it would last that long or take such a horrific toll on human life.

The British Army, or more accurately the British Expeditionary Force (BEF), had landed in Belgium less than two weeks earlier. It had been cautiously moving inland to join up with the French forces when it literally 'bumped in to' the German army – to the surprise of both. The BEF had 70,000 men and 300 artillery guns, whereas the German army consisted of nearly 160,000 men and 600 artillery guns. In hindsight it seems that neither side knew the strength of his opponent but, appropriately, the British took up a defensive position just to the south of the small Belgian mining town of Mons. Its name, like so many others in this area, would become known to the world as a result of the events that occurred in the First and Second World Wars.

This map provides a remarkable picture of the sequence of events in one small sector of the battlefield. To control the 'high ground' has always been considered a military advantage so when the British forces established its defensive line the small hills to the south of Mons were strategically important.

Due to the use of similar colours to represent different features the map initially appears a little confusing but it is possible to decipher it. The roads are marked in brown, the streams and marshy ground in blue, and red contour lines at 20-foot intervals show the actual rise and fall of the ground. As is the norm, the closer the contour lines are together the steeper the slope with Bois la Haut the highest and steepest area shown. The British forces are marked in red and their four initial defensive positions (Posn I, II, III & IV) can be located from north to south. While the troops would have 'dug in', the trench warfare that came to epitomize this war had not yet developed and there was notable mobility on both sides as positions changed dramatically over the day. Careful examination makes it possible to identify the machine guns (MG) and Royal Field Artillery (RFA) as well as the infantry positions and artillery.

The spirits of St George and the Bowmen of Agincourt coming to the aid of the BEF is another facet of the myth of the 'Angel of Mons'.

However, it is the hand-drawn arrows and accompanying annotations that have been added that give such clarity to the numerous actions that took place here and which resulted in some of the heaviest losses of the whole battle. Twenty-four hours later the only British forces on the hill shown here were dead or wounded. Over 400 men of the Middlesex regiment, 300 Irish Guards, many Gordon Highlanders and more than 1,000 German soldiers did not survived the first 24 hours.

Despite the fact that the BEF was forced to retreat from Mons, a 'victory' was claimed because the German 'Schifflen Plan' to sweep through Belgium and northern France into Paris had been halted. 'Victory', as in all conflicts is a relative rather than an absolute term.

The Battle of Mons was also responsible for one of the most extraordinary and persistent myths of the First World War. A tall woman with white flowing robes, sometimes surrounded by bowmen, is said to have glowed in the night sky. The ghostly 'Angel of Mons' was said to have appeared at different points during the battle and to have favoured the British troops. One specific account tells of her rescuing trapped Guardsmen by leading them to a sunken road. Systematic studies have failed to identify any individuals who could say they *definitely* saw her but perhaps this is not surprising. Very few of those who fought here on 23 August were still alive to be interviewed, even months later. Today, the 'Angel of Mons' has become as much a part of the story of the battle as any other events of the day.

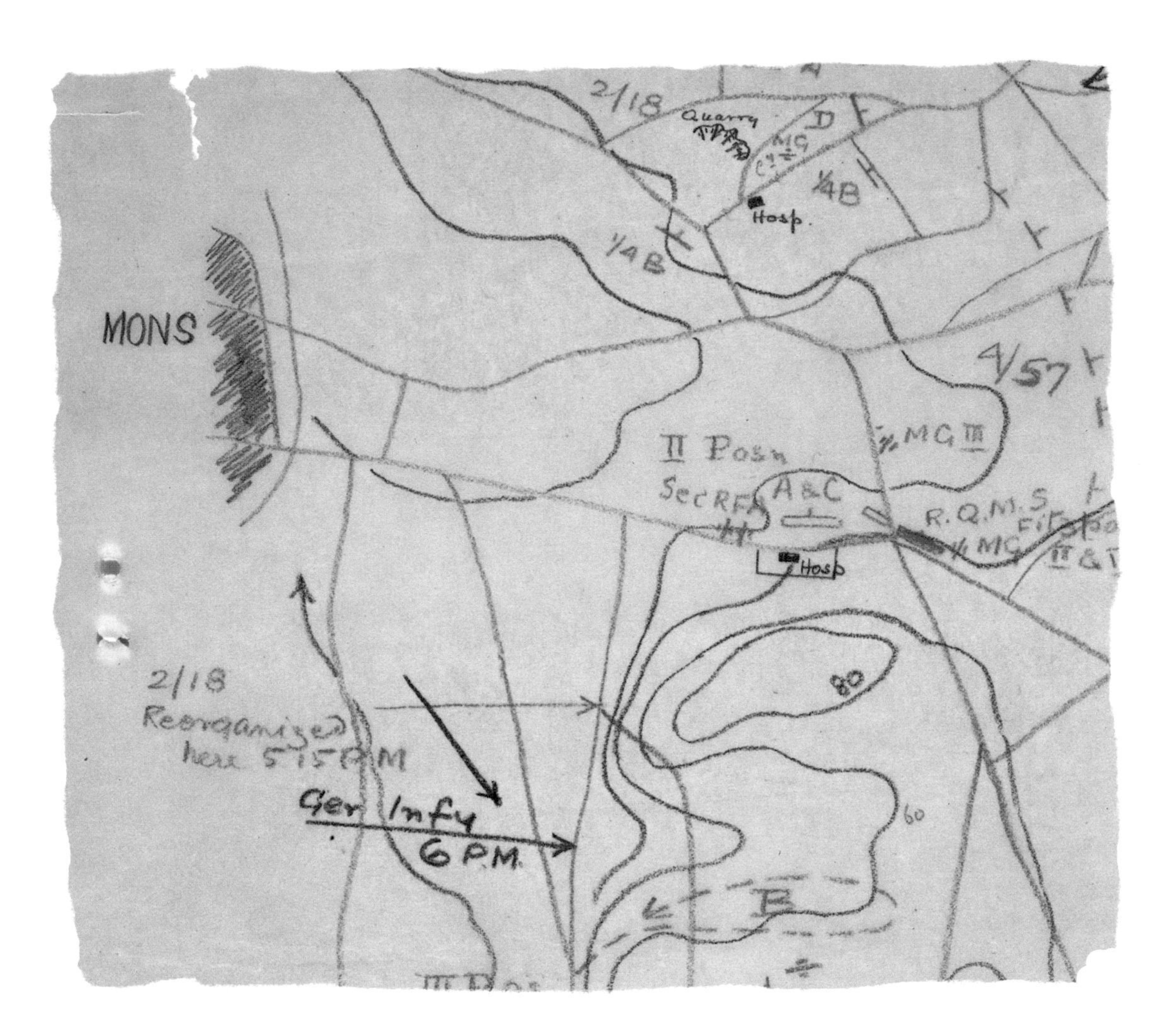

Retreat

There was action, or rather 'actions' all over the area shown. Written records make it possible to appreciate just how much information is actually contained on this sketch map. The inclusion of times also enables one to appreciate just how fast things were happening and how quickly decisions had to be made. And these decisions were literally matters of 'life or death'.

The Germans had attacked relentlessly throughout the day and slowly but surely the men of the 4th Middlesex and Royal Irish Regiments were forced back from their 'I Posn' on to the 'II Posn' shown above, which by late afternoon was also under imminent threat of being overrun. As ammunition supplies ran low and the 'water boiled furiously in the jackets of the machine guns', information was received that the Germans were advancing through Mons. There was a very real chance of being surrounded. As the annotation records, the main body of the Royal Irish Regiment fell back and 'Reorganised here 5.15 PM'. They then provided covering fire for the Middlesex Regiment to retreat through their position and towards Hyon (shown to the bottom left of the main map). However, as the annotation in blue indicates 'Ger Infy 6 PM', when they tried to follow only 45 minutes later a German infantry section had blocked their own route to (relative) safety. From this precarious situation the decision was made to turn north and to try to work their way around the slopes of Bois la Haut and then on to Hyon.

As they did so they came across the 6th Battery of the Royal Field Artillery and their Gordon Highlanders escort in 'IV Posn' (shown below). Not knowing where the next attack or attacks would come from they positioned their guns in a semicircle facing north, south and west just below the ridge of the hill. The men of the Irish now joined this position and dug in as best they could. As a further annotation shows some guns had to be abandoned in the face of the attack from the south. However, whether through insufficient numbers, losses incurred or simply fatigue at the end of the day the Germans did not pursue this attack and the position was held. Then, under cover of dark, this combination of Irish Guards, Gordon Highlanders, Royal Field Artillerymen, and possibly a few others, managed to escape from the hill and join up with the main British forces. While certainly relieved, many of those who survived this first day would be killed over the coming week as the numerically superior German forces harassed the retreating British.

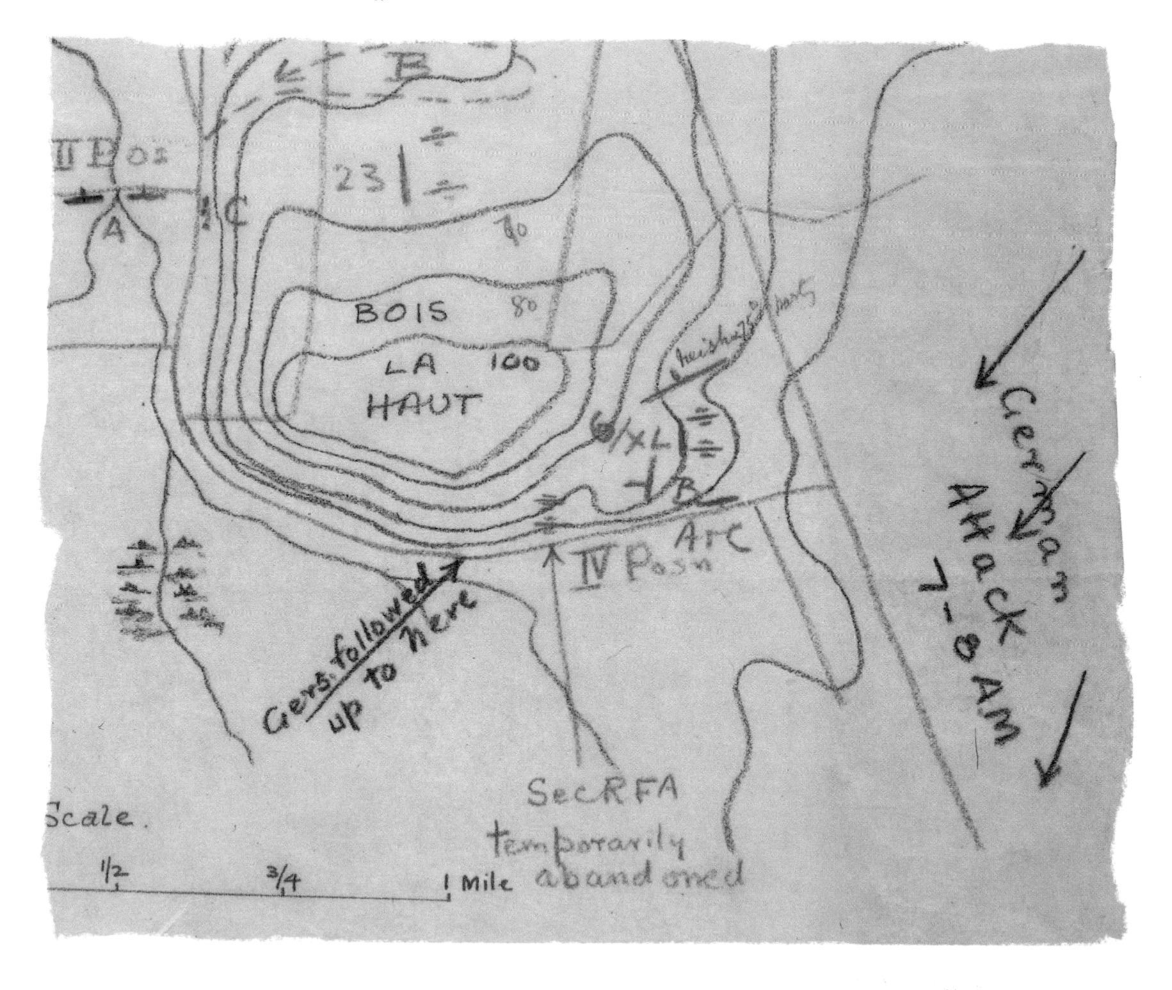

BIGOT
TOP SECRET
A.S.V.
NIGHT
ANTI
'U' BOAT
ANTI
SURFACE
CHERBOURG
MAISY
BAYEUX
CAEN
LE HAVRE
OUISTREHAM
LISIEUX
ST. LO
EVREU
ARGENTAN
DOMFRONT
DREU
ALENCON
RENNES
LAVAL
LE MANS
CHATEAU
TAC/RECCE
P. R.

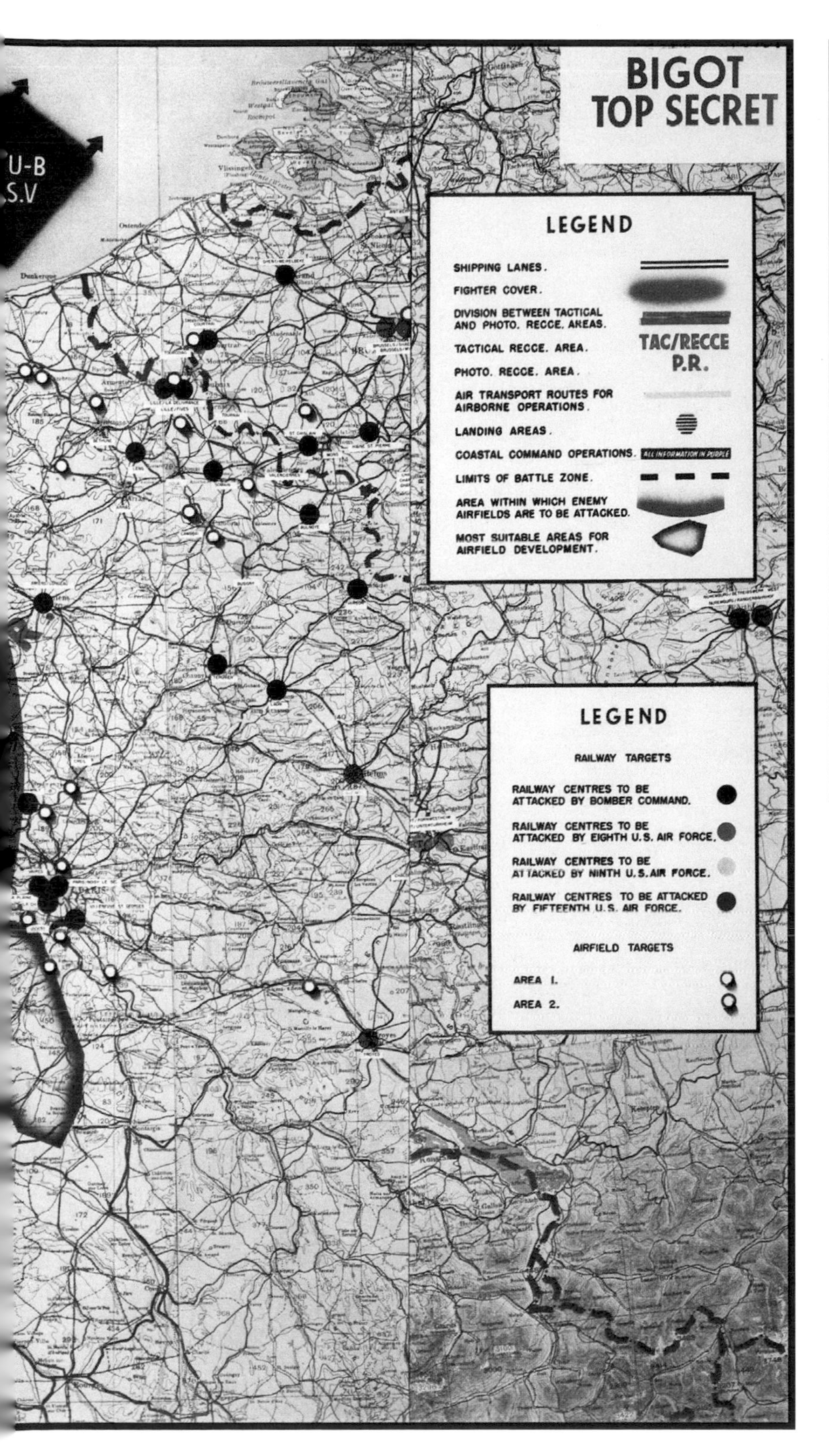

BIGOT
TOP SECRET
U-B
S.V
LEGEND
SHIPPING LANES.
FIGHTER COVER.
DIVISION BETWEEN TACTICAL AND PHOTO. RECCE. AREAS.
TACTICAL RECCE. AREA.
PHOTO. RECCE. AREA.
TAC/RECCE
P.R.
AIR TRANSPORT ROUTES FOR AIRBORNE OPERATIONS.
LANDING AREAS.
COASTAL COMMAND OPERATIONS.
ALL INFORMATION IN PURPLE
LIMITS OF BATTLE ZONE.
AREA WITHIN WHICH ENEMY AIRFIELDS ARE TO BE ATTACKED.
MOST SUITABLE AREAS FOR AIRFIELD DEVELOPMENT.
LEGEND
RAILWAY TARGETS
RAILWAY CENTRES TO BE ATTACKED BY BOMBER COMMAND.
RAILWAY CENTRES TO BE ATTACKED BY EIGHTH U.S. AIR FORCE.
RAILWAY CENTRES TO BE ATTACKED BY NINTH U.S. AIR FORCE.
RAILWAY CENTRES TO BE ATTACKED BY FIFTEENTH U.S. AIR FORCE.
AIRFIELD TARGETS
AREA 1.
AREA 2.

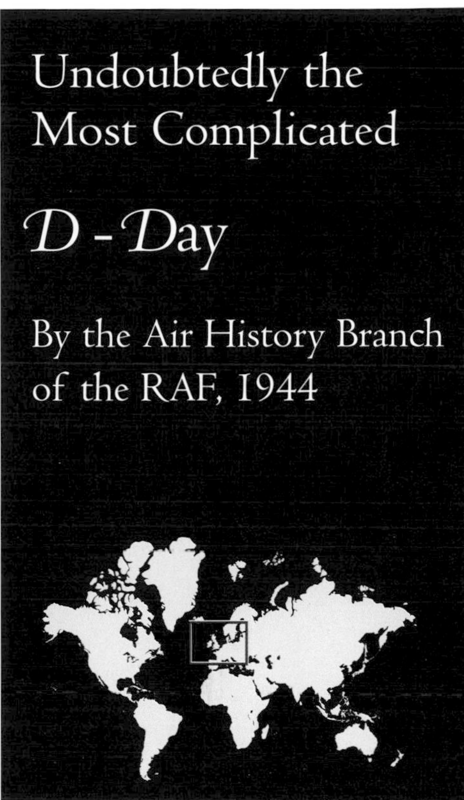

Undoubtedly the Most Complicated
D - Day
By the Air History Branch of the RAF, 1944

Undoubtedly the Most Complicated

D-Day

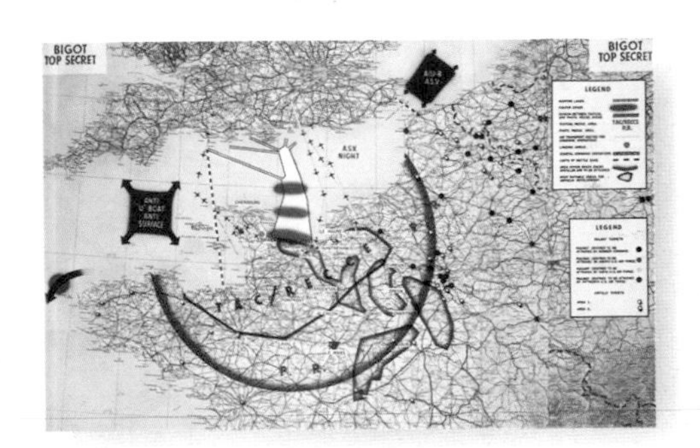

D-Day: 6 June 1944. Some 60 years after this date the phrase 'D-Day' continues to be used as a colloquial term to describe 'a make or break' course of action. It is an accurate and appropriate interpretation of what happened and what was at stake on that summer's day.

The invasion fleet lies at the heart of the story of D-Day. With around 6,000 vessels, ranging from battleships to small landing craft, ferrying 150,000 soldiers across a choppy sea in the dark to face an uncompromising enemy, Churchill's description of it as 'undoubtedly the most complicated and difficult' military undertaking in history is not an exaggeration. While tales of heroism and fortitude on the ground are rightly shared and celebrated, the success of the invasion was based on meticulous planning and preparation. Coordination was probably the biggest challenge of all. The multi-national composition of the forces posed its own difficulties, not least in terms of conflicting personalities. Beyond this, success depended upon unprecedented levels of cooperation between the three branches of the military. And it all had to be done in secret. The risk that the Germans would find out when and where the invasion would take place was the greatest fear. The 'need to know' principle was applied rigorously and while individual men, units, ship and air crews were all trained in relation to their particular objectives, only a tiny number at the very top had an overview of the whole.

Written communications could be encoded but maps were difficult to disguise and remain useful and were thus particularly vulnerable – yet they were crucial to the planning process. Any of the individual details included on this map would have given away the true intentions of the Allies. Created by the Air History Branch of the RAF it details the crucial role played by the RAF, alongside the USAF, in the days of early June 1944. Although this particular map focuses on its role in Normandy it needs to be understood that the RAF continued to undertake numerous raids and reconnaissance missions further north and on the Mediterranean coast to confuse the enemy. The bombing of the

Aerial photograph of a temporary harbour formed by the Allies on the Normandy coast shows the breakwater made up of merchantmen and old warships.

railway centres and airfields inland from Calais, marked here, would have been intended to both support the German belief that this is where an invasion would occur while also seeking to ensure that neither enemy aircraft nor army divisions could play a part in what was to happen further down the coast.

However, what this map does clearly illustrate is that this was simply one aspect of the diverse roles played by the RAF.

From the moment the decision was made in 1943 to invade France the top priority was to take control of the skies and to incapacitate the Luftwaffe. While seeking to maintain the deception by spreading their raids, the sweeping curved red line shows those airfields within easy reach of the invasion beaches were specifically targeted.

Alongside the destructive role played by the opposing air forces, their role as 'eyes' or information providers was at least as important. As the Luftwaffe's effectiveness decreased so it denied Hitler's High Command the visual information that might well have given away the location of an army of over 1,000,000 and an invasion fleet of 6,000 vessels! As the markings in green and blue on this map show, the RAF's role in undertaking tactical and photographic reconnaissance was given the highest priority.

The actual routes of the airborne missions shown by the yellow lines are likely to have been informed by some of the aerial reconnaissance information. For example, the selection of the targets for the paratroopers and glider divisions near to Ouistreham, which was crucial to the success of the main invasion force, was a result of intelligence gained from the air. Even so, it was not infallible, as the paratroopers who landed in areas that had been deliberately flooded found to their cost.

The invasion fleet's route is shown in white in the centre of the map and the hazy red shading above it indicates the air cover provided by the Spitfires, Hurricanes and Mosquitoes. While the initial force travelled through the night of 6 June unmolested the subsequent transfer of the main body of the army and its supplies over the next few weeks called this cover into practice on many occasions.

And yet even the major and diverse roles identified on this map do not tell the whole story. There is, for example, nothing in the key to show where small single planes dropped agents or landed supplies for the resistance in occupied areas.

This map provides a fascinating insight into the many roles undertaken by the RAF and perhaps also enables a deeper appreciation of the contribution made by those coordinating Operation Overlord in 1944.

Those Guns Must be Silenced

Ouistreham is a small port at the seaward end of a canal that leads to Caen. On 6 June 1944 it was of supreme strategic importance. The yellow lines with the small planes mark the route of the planes and gliders carrying the men of the Parachute and Air Landing Brigades. They were the first troops to land as part of the invasion, and the fate of the main task force, which was just setting sail as they dropped from the skies, to a large part depended on their success. The destruction of most of the bridges over the canal would make it impossible for the German tank divisions further south to arrive quickly, while the capture and defence of 'Pegasus Bridge' (as it became known after the cap badge of the 'Paras') would allow the Allies to move across this obstacle as required. Less

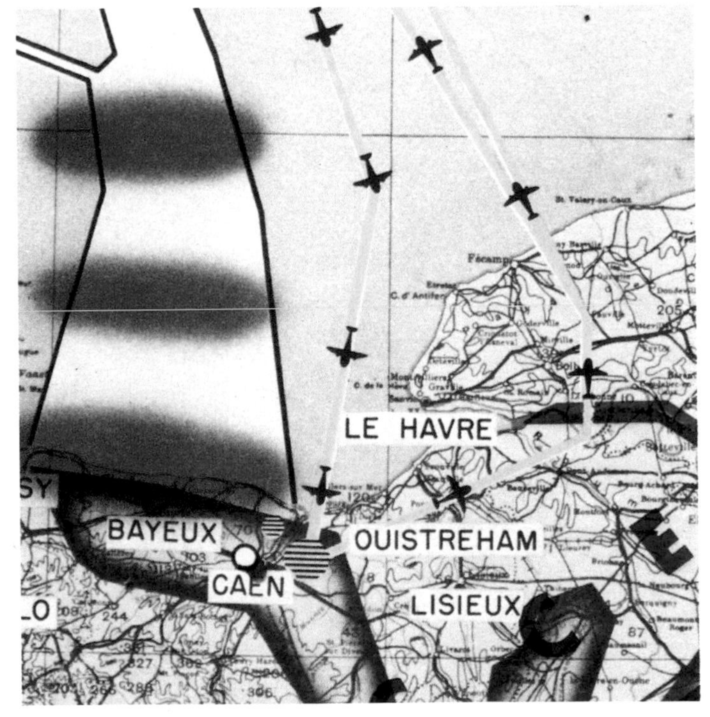

well known, but just as significant, was the capture of the major gun emplacement in the area. The Merville Battery was a massive and heavily fortified gun emplacement that guarded both the sea approaches to the canal and the flat sandy beach to the south; that is, the route of the incoming allied task force and Sword beach in particular. Such was its concrete covering, some 3.5 metres (12 feet) thick, that several allied air attacks had failed to put it out of action. The airborne attack had been practised many times in England but almost nothing went to plan on the night. Nevertheless, at incredible cost the battery was captured before the first allied ships came over the horizon. It was to change hands several times over the next 24 hours as the battle ebbed and flowed but its guns played no part.

The neatly drawn yellow lines shown here with their precise 'out this way', 'home this way' belie the accidents, misfortune and necessary improvisation that occurred once the first planes took off. The skills and bravery of the glider pilots, the tug (or towing plane) pilots and of the planes carrying the parachutists were tested to the full.

The E-boat Threat

The two black shapes on the map record another significant way in which the RAF contributed to the success of D-Day before, during and after 6 June. The massive, closely grouped and slow-moving invasion fleet would have been the easiest of targets for both enemy submarines and the fast E-boat (motor torpedo boat) flotillas based in the channel ports. As the map shows, the RAF (Coastal Command) was allocated the role of establishing and maintaining a 'no go' zone between the North Sea and the Atlantic south of Cornwall. Over 2,000 flights were made, some 70 submarines sighted and 40 attacked and the threat was neutralized. By maintaining a strong presence at both ends of the possible invasion zone no clues were given as to the intended location of any invasion.

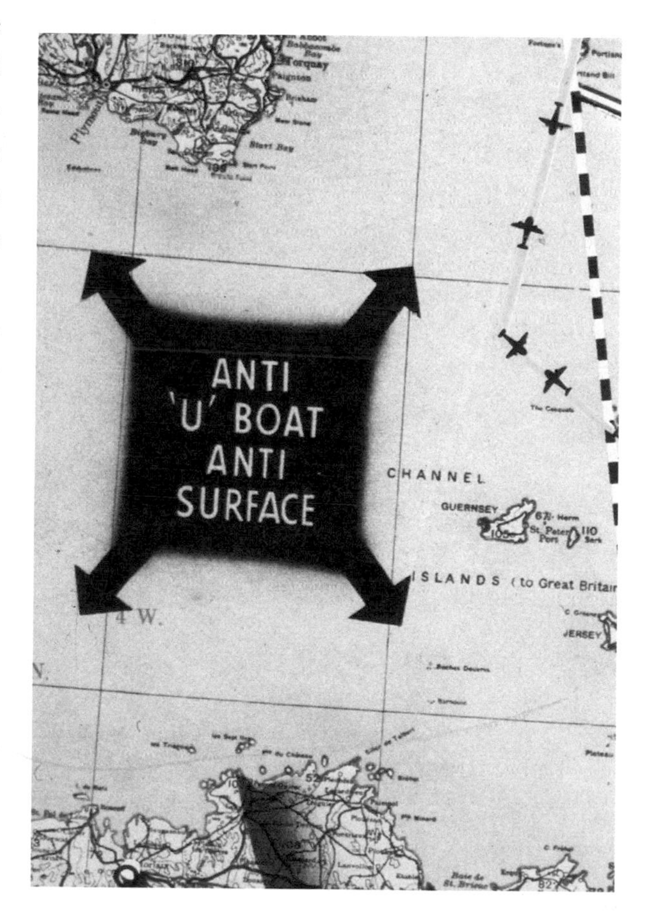

Once the liberation of Europe had begun, an early target was the large E-Boat depot at Le Havre, which was effectively destroyed by heavy bombing one week after D-Day. To have attempted to do so prior to 6 June would have risked losing the element of surprise. However, the week in-between was a key one but the combined operations of RAF Coastal Command and the Royal Navy prevented the E-boats causing any serious disruption to the ships now making their way in both directions between England and France.

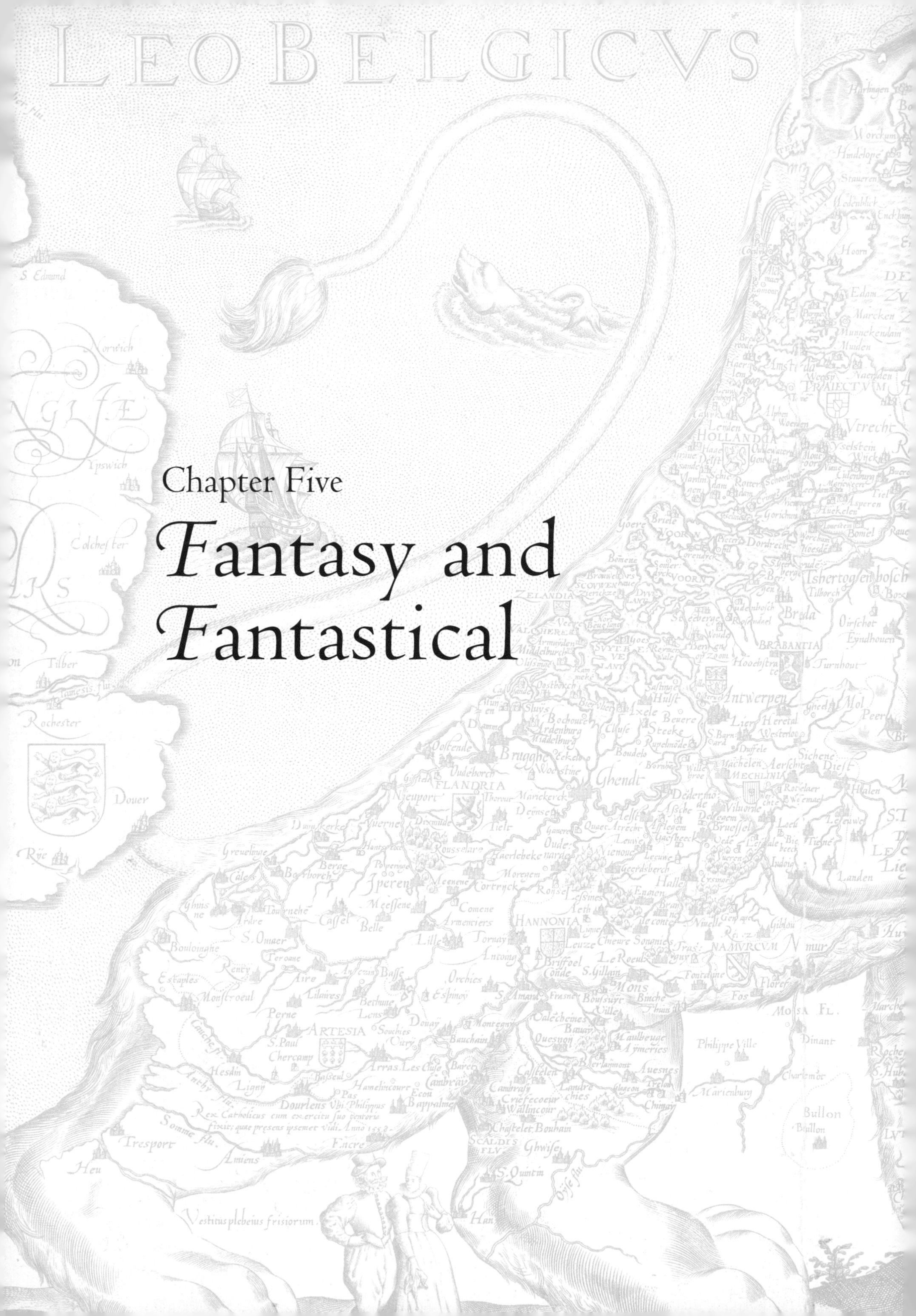

Chapter Five

Fantasy and Fantastical

LONDON
S V R R E Y
S·V·S·S·E·X
HASTINGS
CHICHESTER

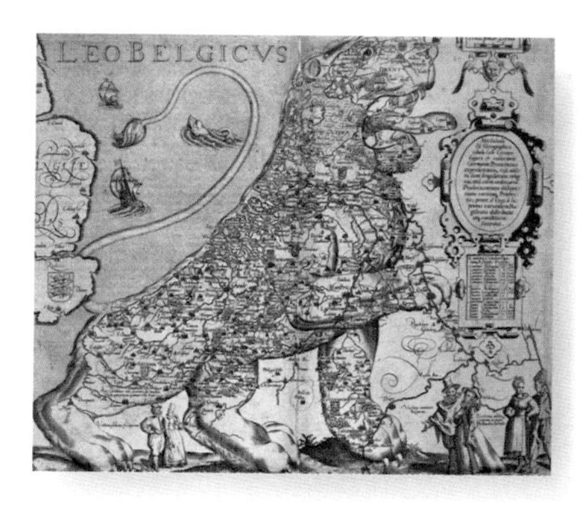
LEO BELGICVS

NUEVO
MEXICO
CANADA
NOUVEL
LE FRANCE
NUE VA
GRA NA
DA
MER
DE
SUD
PACIFICQUE
MEXICQUE

PARADISE OR THE GARDEN OF EDEN

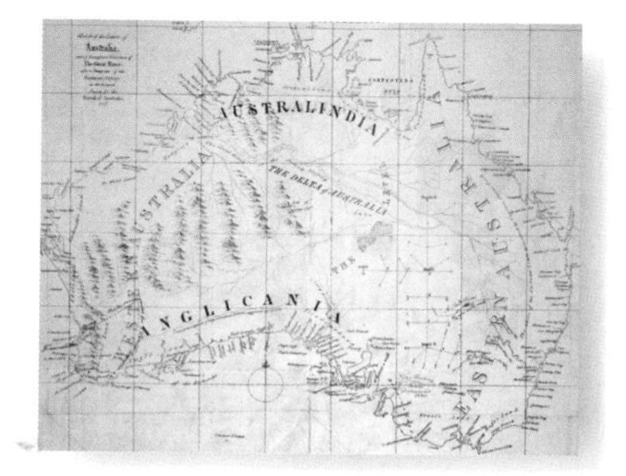
AUSTRALINDIA
ANGLICANIA

A map is an incredibly flexible and adaptable medium. As has been illustrated in the previous chapters it can be of any size, be coloured or plain, be illustrated, annotated and modified. What all maps share is the intention to record and communicate 'something'. An ever-present and underlying premise is that if it is on the map then it does exist.

There are clearly many fictional exceptions, such as the 100 Acre Wood where Winnie the Pooh and his friends live or the Kingdom of Middle Earth where the action in the *Lord of the Rings* takes place. It is interesting to reflect on how one's understanding of the text is enhanced by access to this visual presentation of the locations and the spatial relationship between them. One such literary example is included in this chapter. The maps used by the poet Michael Drayton in his epic *Poly Olbion* (pp. 156–61) were forerunners of these modern illustrative maps.

Beyond this literary sphere the 'if it is on the map it exists' assumption has given and continues to invest maps with a certain authority. The stories behind the maps included in this chapter show how supposition and hearsay can give rise to the belief in geographical realities that are not rooted in fact.

This can perhaps be best seen in relation to the map of California (pp. 168–73). It is shown on Nicholas Sanson's map as an island but this is clearly inaccurate because we *know* this is not the case. But it is interesting to pause and reflect on how we know this and feel so confident about this knowledge. Very few readers will have walked out on to the peninsula even if they have visited this part of the world. It is highly likely that our knowledge comes from the maps we have grown familiar with, possibly supported by aerial photographs or satellite images. We are actually making the same assumptions about the accuracy and validity of the information that have been made through history.

What the California map does is remind us that a map is a reflection of the existing state of knowledge. At the present time the mapping of the deep ocean beds and the distant galaxies are comparable to the mapping of the world that took place in the past. Both are incomplete, are cumulative and will contain errors that will be 'corrected' in the future. Interestingly, explanations of maps of the seabed and outer space seem to always include a significant element of interpretation of the available data to 'fill in the gaps'. Future generations may view our current maps of these locations in a similar manner to the way we view 'Island California' or the 'Great River' (pp. 180–85) today.

It would also be naive to view the development of the maps as linear and with each new one being in some way superior to those that preceded it. The map of the Garden of Eden (pp. 174–9) undoubtedly possesses a certain period charm but may prove to be more geographically accurate than it has been given credit for. Today the internet contains a large number of sites relating to ongoing attempts to locate the Garden, particularly using satellite technology to establish the course of the rivers identified as flowing from it. Other sites show the same technology being used in conjunction with maps like this one to find the final resting place of Noah's Ark. It is not inconceivable that these locations may in the future be plotted on maps once again.

El Dorado, or the City of Gold, also continues to attract the active attention of some people in the twenty-first century. When one considers the incredibly rich finds from the Ancient World that continue to be discovered on a regular basis who is to say that the stories relating to a City of Gold will not one day prove to be founded and, again, be 'added to the map'? Many years ago the city of Troy as featured in Homer's *Iliad* was believed to be 'mythical' but archaeological work has tentatively established an actual location and it is 'on the map' once again.

The political situation in all parts of the world is never static and one map included in the section illustrates this dynamic clearly. To present a map of the Netherlands, Belgium or Luxembourg in the shape of lion would have no relevance today. Their status as established independent countries is beyond question and the era when they were provinces of a colonial empire under a Spanish king has long passed. But at the time the *Leo Belgicus* map (pp. 162–7) was produced the propaganda value of its symbolically defiant shape was enormous.

Adam and Eve's expulsion from the Garden of Eden. The search for its location using clues from the Bible continues.

However, it can be quite confidently predicted that at least one of the maps included in this chapter is unlikely to ever have its information confirmed or to be updated. The eccentric maps of the regions of England and Wales in the *Poly Olbion* are inextricably linked to the poem they were created to accompany. Hills remonstrating with rivers as they engage in flirtatious relationships with each other while forests complain about their fate is not the usual way they are represented on maps. Nevertheless, the combination of poetry and cartography employed by the poet prove to be mutually supportive and *Poly Olbion* would be a less impressive work without its maps. Although it did not achieve the popular acclaim its author felt it merited it remains a particularly distinctive and early example of the versatility of maps.

Maps are indeed a very adaptable and powerful form of communication. The following illustrate how they can intrigue, inspire and mislead: sometimes all at the same time!

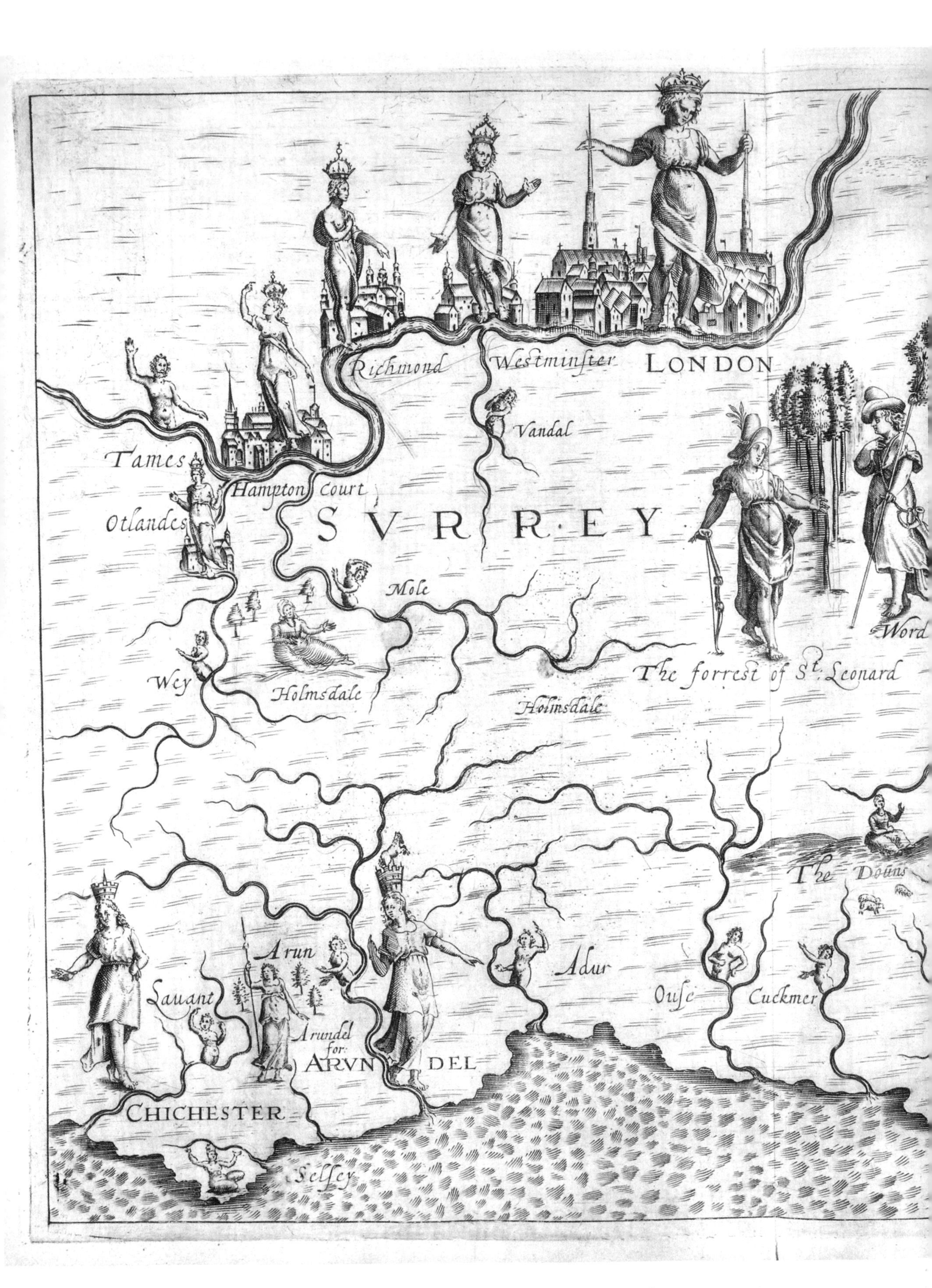

Richmond
Westminster
LONDON
Vandal
Tames
Hampton court
Otlandes
SVRREY
Mole
Wey
Holmsdale
Holmsdale
The forrest of St. Leonard
Word
The Douns
Arun
Lauant
Arundel for:
ARVN DEL
Adur
Ouse
Cuckmer
CHICHESTER
Selsey

This Blessed Land

Southern England from the Poly Olbion

By William Hole, 1612

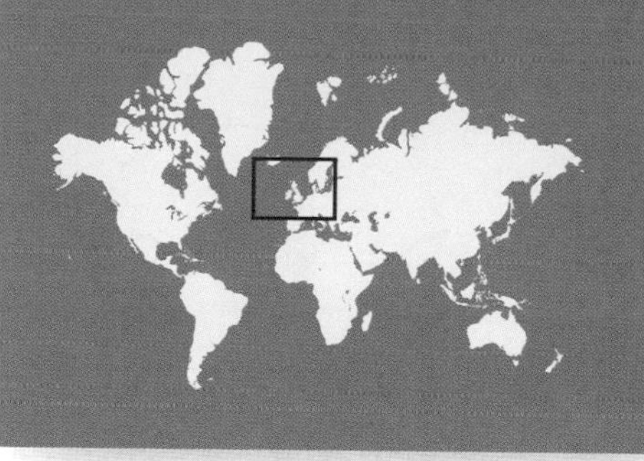

This Blessed Land

Southern England from the *Poly Olbion*

It is said that for every successful writer, poet or artist, there are hundreds who 'aspired' but never realized their ambitions. This unusual map forms part of the story of one such poet. While any discussion of the golden literary period of the late Elizabethan and Stuart period will inevitably lead to references to William Shakespeare and Christopher Marlowe the name of Michael Drayton (1563–1631) is unlikely to feature. Yet Drayton was a contemporary poet, respected by his peers and confident that he was the equal of any of them. After a number of smaller works he set about creating what he believed would be an epic masterpiece that would finally establish him among the elite. He began work on this in 1598 and the first edition was published in 1612. Its title was *Poly Olbion*, which is generally translated as 'Britain: Blessed in Many Ways'. This was an idiosyncratic work. Drayton created a wandering Muse who visits each part of the country and recounts chosen events about the locality through a 'song'. What is so unusual in this work is Drayton's use of each county's physical geography – its rivers, hills and valleys – through which to tell his tales. His epic tome was also lavishly illustrated with a map to accompany each section, or 'song' as he called them. This particular example features the southern counties of Surrey and Sussex. As can be seen, however, this is no traditional topographical map for the rivers, hills and forests are given human form through which to tell their tales as Drayton makes full use of the poetic device of personification. The whole has a nationalistic, celebratory tone as those living in 'this blessed land' overcome all trials and tribulations.

The evocative maps are a key part of *Poly Olbion* and were created by William Hole (d. 1624), one of the leading cartographic engravers of this time. Alongside the verse and the maps, Drayton's tome also included annotated notes by William Selden, a leading historian of the day. It was clearly intended to both amuse and inform. Despite the 15 years of his life Drayton invested in this work, despite its rich and lavish verse, despite its unique maps and historical commentary, the *Poly Olbion* was not a success. Drayton persevered with a further 12 sections and published an extended second edition several years later. This met with no greater response than the first, much to the poet's bewilderment and exasperation.

Today it is viewed as a distinctive example of the late Elizabethan/early Jacobean period and has captured the attention of literary scholars. Drayton has become something of a cult figure in certain circles and his verse continues to be scrutinized for hidden political meaning reflecting the changes under way in his society. Even so, the remarkable and distinctive maps are probably more widely known than the text they were drawn to illustrate.

On this map we have the Muse arriving in southern England. The style is indicative with rivers forming the dominant feature and very few towns being shown. The quintessential nature of *Poly Olbion* can be best appreciated by examining the central section where the link between the verse and the map is made. There are four larger figures with a number of smaller ones below. The larger figures represent the forests of Surrey and Sussex – St Leonard, Word, Ashdowne, Waterdowne (plus Dallington close to the coast) – and the somewhat sombre posture adopted is significant. They are complaining to the Muse about the abuse and mistreatment they are being subjected to which is causing their size to diminish. Their complaints serve only to amuse the hills to the south (The Downs) who benefit from having the trees cleared from their slopes. Two smaller sitting figures represent these hills and a closer look reveals that they seem to be laughing and waving their hands dismissively at the forests' complaints. But there is more. The animated gestures of tiny female figures arising from the nearby streams are remonstrating with The Downs for their selfish response to the plight of the forests. The Muse attempts to sooth the situation before continuing the journey into Kent and the wonders found there. Unfortunately Drayton's idiosyncratic approach failed to capture the imagination of the public.

Drayton's poem was an epic journey through Britain's countryside, encountering woods and streams, hills and rivers.

However, his name does form part of a popular legend involving his more celebrated contemporary, William Shakespeare. There is a story that the fever which caused Shakespeare's death in 1616 was precipitated by a heavy drinking session with friends, one of whom was Michael Drayton!

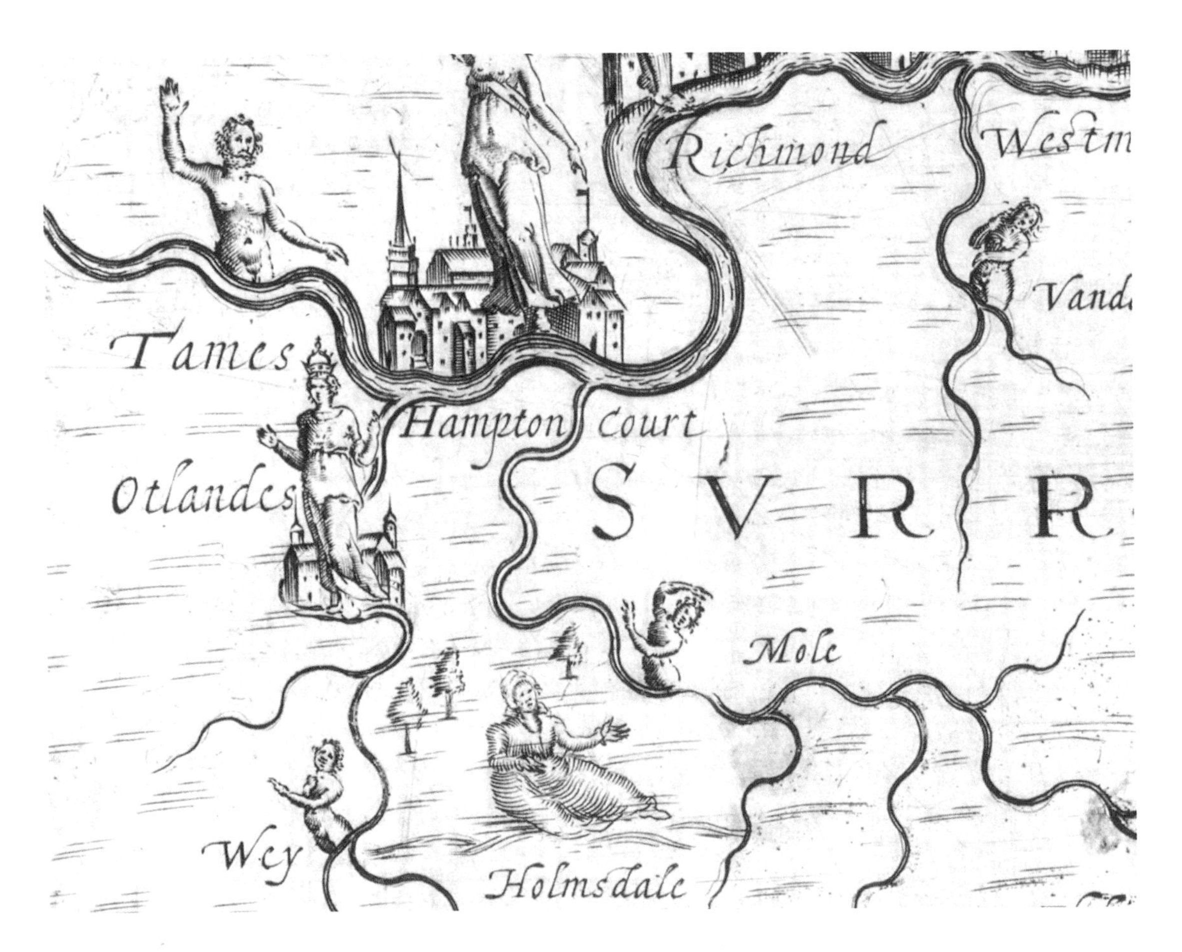

Passion Runs Deep

Today, as in the sixteenth century, the River Mole flows into the River Thames on the opposite bank to Hampton Court Palace. In Drayton's epic this is not regarded as a simple geographical feature but a dramatic and passionate event. The bearded and masculine Father Thames is shown waving and the object of his attention is the young female that is the River Mole. Her 'body language' is intended to convey flustered but irresistible attraction. In Song 17 of the *Poly Olbion* the poet recounts how the affair between the Thames and the Mole is one neither could deny, such was the appeal. However, this liaison did not meet with the approval of either set of 'parents'– these 'parents' being the hills that provided the waters for their 'offspring'. The image and posture adopted by the motherly figure of Holmsdale (hills) is in the classic pose of a parent attempting to reason with an errant child who, as here, is oblivious to her entreaties.

This theme of river convergence being presented as an affair or a marriage is one that Drayton uses elsewhere. When his Muse visits Oxfordshire Drayton documents an elaborate 'marriage' of the Rivers Isis and the Tame before ending that 'The wedding ends, [the Tames] got, borne, and bred, immediately doth flowe to Windsor-ward'.

As is shown here, however, Drayton's metaphorical marriages seem to have been just as susceptible to temptation as those in the real world!

Spires Rising Over London

London dominates the map with the Goddess looking down benignly on the city below. She stands on the banks of the Thames holding one tall church steeple tightly and seeming to steady herself on the other. These would seem to be the spires of London's two major religious buildings, Westminster Abbey and St Paul's Cathedral. However, the steeple of St Paul's (on the right) had been destroyed in a fire in 1561, some 50 years previously. Although it had been the tallest structure in the whole of Britain, it was never replaced. Drayton's inclusion of it here is just a further small example of his nostalgia for the grandeur of the Tudor period which, even as he started on his major work, was drawing to a close.

Drayton would not have known, though he may have dreamed, that the other steeple would one day rise above his final resting place. When he died in 1631 he was buried in Westminster Abbey. While a bust of him can be found in Poet's Corner he was actually buried under the north wall of the knave and, in a way, this 'so near yet so far' reflects his life and his claim to be a 'major' poet of his time. As his work, and in particular the *Poly Olbion*, continues to attract critical appreciation in modern times it is not inconceivable that he might, one day, be laid alongside his more famous contemporaries in Poet's Corner. His Muse would surely have deemed this appropriate.

LEO BELGICVS
22
23
24
25
26
27
53
52
51
50
Hul
Hüber flu.
Hable
S Edmund
Norwich
ANGLIÆ
Ypswich
PARS
Colchester
London
Tilber
Thamesis flu.
Rochester
Douer
Rye
Leeuwaerden
Harlingen
Bolswaert
Franicker
Worckum
Hindelope
Stauoren
Slooten
FRISLAN
Steenwick
Medenblick
Enckhuysen
Hoorn
Vollenhoue
DE Campen
ZVYDER ZEE
Edam
Marcken
Munnekendam
Elburg
Harderwyck
Amstredam
Naerden
Elsbeet
ZVTPHAN
Emmers voort
PRAIECTVM
GELDRIA
Vtrecht
HOLLANDIA
Leyden
Woerden
Rhenen
Arnehem
Bronchorst
Doesburch
Nieumegen
Batenburg
Gorichom
Tiel
Asperen
Leuwe
Bomel
Heusden
Goere
Briele
VOORN
Dordrecht
Graue
Cuyck
Tshertogenbosch
Tilborch
Bosteel
Breda
ZELANDIA
WALCHERE
Middelburch
Vlissinge
Berghen op Zoom
Rosendael
Oirschot
Eyndhouen
Helmont
BRABANTIA
Hoochstrate
Turnhout
Antwerpen
Hulst
Axele
Beuere
Steeke
Lier
Heretal
Westerloo
Mol
Peer
Weert
Hoorn
Wessen
Maes Eyck
Brey
Stochem
Sluys
Ardenburg
Damme
Ostende
Brugghe
Oudeborch
FLANDRIA
Ghendt
Nieuport
Dixmude
Duynkerke
Vuerne
Berge
Bourborch
Greueling
Cale
Rousselare
Tielt
Deynse
Ipperen
Meenene
Cortryck
Harlebeke
Oudenarde
Aelst
Mechelen
MECHLINIA
Aerschot
Diest
Sichene
Maestricht
Valke
Vilvorde
Bruessel
Louen
Tienen
S.Truden
Tongeren
LEODIVM
Liege
Landen
LIMBVRG
Limburg
Hall
Ath
Enghien
Nivelle
Gemblou
Giblou
Huy
Spa
Ghisne
Ardre
S.Omaer
Cassel
Belle
Mesene
Comene
Armentiers
Lille
Tornay
HANNONIA
Leuze
Mons
LeRoeu
NAMVRCVM
Namur
Bouloinghe
Teroane
Renty
Aire
Lilaires
Basse
Bethune
Lens
Orchies
Antoing
Conde
S.Gillain
Binche
Florefe
Estaples
Monstroeul
Perue
ARTESIA
Souchies
Douay
Valencyn
Bouchain
Quesnoy
Maubeuge
Auesnes
Philippe Ville
Dinant
MOSA FL.
Marche en Famines
La Roche
Durbuy
Staueloo
S.Vit
Cleiff
Rochefort
S.Hubert
Bastogne
S.Paul
Chercamp
Hesdin
Ligny
Arras
Cambray
Bapalme
Cateau Cambresis
Landrechies
Chimay
Marienburg
Charlemont
Bullon
LVTZENBVRG
Diekirch
Arlon
Luxembourg
Virton
Longwy
Villers
Thionville
Dourlens
Vbi Philippus Rex Catholicus cum exercitu suo ventoris fixit; quae presens ipsemet vidi Anno 1558
Somme flu.
Tresport
Amiens
Encre
Heu
Chastelet
Bouhain
SCALDIS FLV
Ghwise
S.Quintin
Oyse flu.
Han
Vestitus plebeius frisiorum.
Paris
P. Kaerius ex.

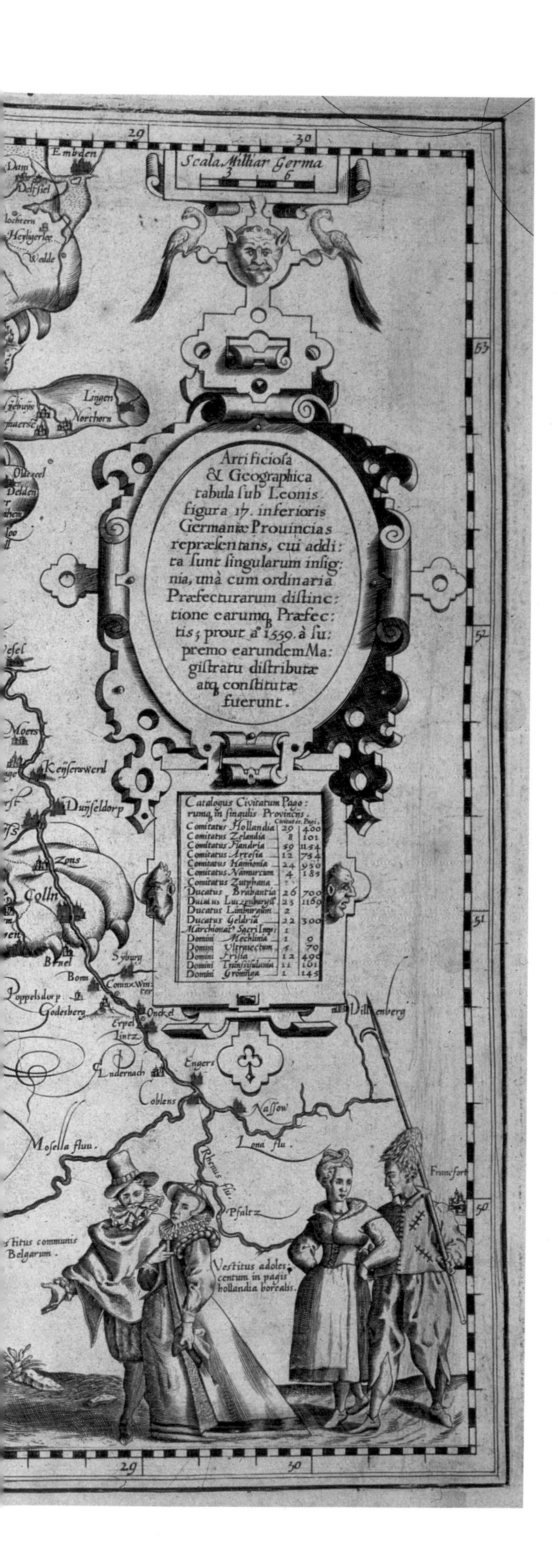
Scala Milliar Germa
Embden
Artificiosa
& Geographica
tabula sub Leonis
figura 17. inferioris
Germaniæ Prouincias
repræsentans, cui addi:
ta sunt singularum insig:
nia, unà cum ordinaria
Præfecturarum distinc:
tione earumq Præfec:
tis; prout aº 1559. à su:
premo earundemMa:
gistratu distributæ
atq constitutæ
fuerunt.
Catalogus Civitatum Pago:
rumq in singulis Provinciis
Comitatus Hollandia 20 400
Comitatus Zelandia 8 101
Comitatus Flandria 59 1154
Comitatus Artesia 12 754
Comitatus Hannonia 24 950
Comitatus Namurcum 4 183
Comitatus Zutphana
Ducatus Brabantia 26 700
Ducatus Luxenburgii 23 1169
Ducatus Limburgum 2
Ducatus Geldria 22 300
Marchionat' Sacri Imp. 1
Domini Mechlinia 1 0
Domini Ultrajectum 5 70
Domini Frisia 12 490
Domini Transisulania 11 101
Domini Groninga 1 145
Lingen
Northorn
Moers
Keysserswerd
Duysseldorp
Zons
Colln
Sybburg
Bonn
Poppelsdorp
Godesberg
Onckel
Dillenberg
Lintz
Engers
Andernach
Coblens
Nassow
Mosella fluu.
Lona flu.
Rhenus flu.
Pfaltz
Francfort
Vestitus adolescentum in pagis hollandia borealis.
Belgarum.
29
30

Roaring Defiance
Leo Belgicus
By Pieter van den Keere,
1617

Roaring Defiance

Leo Belgicus

This 1617 map showing a roaring lion with its head back, foot raised and claws extended, presents a defiant image – and this is deliberate. The 'Lion map' of the Low Countries (present-day Belgium, the Netherlands and Luxembourg) first appeared in a book written by an Austrian nobleman in 1583. Michael Aitzinger's imagination had been sparked by the overall shape of the land and probably also by the fact that all 17 provinces incorporated a lion in their coats of arms (as shown by the small shields on the map). He would surely have been amused at his novel representation and pleased with the positive reception it received.

As his *Leo Belgicus* (Belgian Lion) concept was copied and adapted by subsequent mapmakers its 'defiant' image acquired increasing significance. In the sixteenth century the countries of Belgium, the Netherlands and Luxembourg did not exist and the United Provinces as they were then known formed part of the Spanish Empire. As such, this part of the world was caught up in the political and religious conflicts that dominated those times. Spain was the richest and most powerful country in Europe, and its King, Phillip II, believed that it was his duty to defend Catholicism against the emergent Protestantism. The Protestant faith had established itself in the cities of the northern provinces and despite the presence of a large Spanish army and persecution by the infamous Spanish Inquisition, a war of independence broke out.

Throughout the 1580s the Spanish army's success against the Protestant rebels, or freedom fighters, prompted Elizabeth I of England to send support. Incensed, Phillip sent the legendary Spanish Armada to conquer England. Had the victory of this famous sea battle fallen to the Spanish the history of the countries that face on to the North Sea would have been quite different. Instead, the Spanish hold on the Low Countries was seriously weakened by this defeat and the Dutch achieved independence in 1648. However, it was not until 1839 that all three former provinces finally achieved independent nation status. This did not prevent each continuing to be at the centre of the European and world conflicts that raged over the following century.

This version of the *Leo Belgicus* was created in 1617 by Pieter van den Keere (or Kaerius) (1571–*c.* 1646), who had had to flee to England to escape the religious persecution in 1584. It was in London that he met a fellow refugee and compatriot, Jodocus Hondius, from whom he developed his skills of engraving and cartography. He worked in London for many years and produced many celebrated English county maps. Although there were a number of *Leo Belgicus* maps produced either side of 1700, his was one of the most elaborate. However, its 'novelty' should not be allowed to cloud its value as a map in its own right. Cities and major towns are accurately located while the chart (on the right side of the map) provides details of the number of towns and villages in each province. What makes this version important is the manner in which the political situation at the time is portrayed by the colouring of the towns. In the inside cover of the atlas of which this map forms a part there is a longhand written key which has obviously been

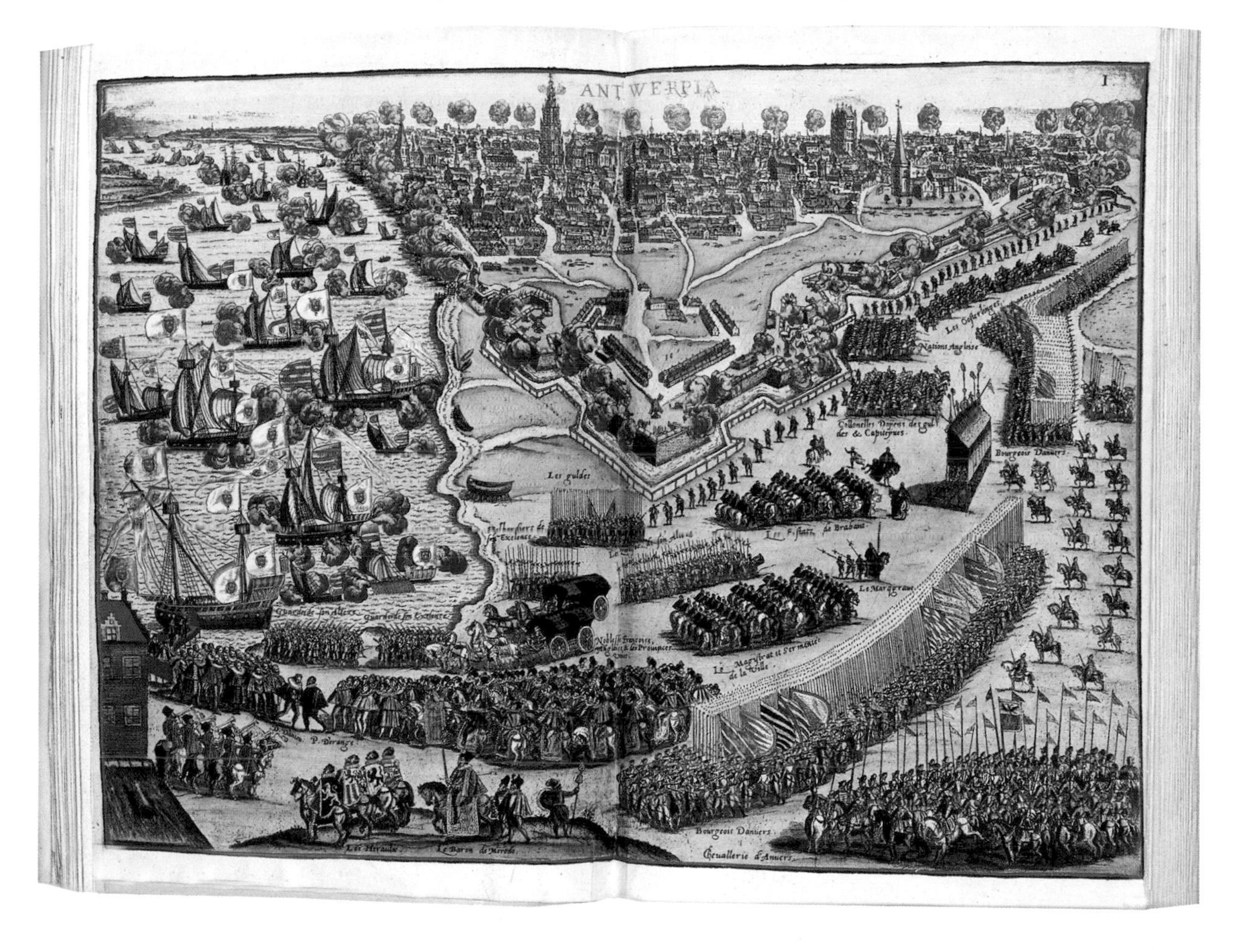

Antwerp's strategic position on the River Scheldt meant that it was at the centre of the struggle between the Spanish and the Dutch.

added after publication. In a spidery hand it explains that red is for 'the King of Spain', yellow for 'the states', and the smaller number of 'blew' for 'the Emperialls' and green for 'the King of Frances'. Whether van den Keere or the purchaser added this explanatory annotation is uncertain but it was clearly considered very important to record who held what at this time. The ongoing nature of the conflict is really captured by those towns that have been given a blue or yellow stroke alongside the main colour. As the key informs us, these were towns in the possession of either the Spanish (red stroke) or the states (yellow stroke) but whose destiny had yet to be decided.

Not surprisingly considering the leadership provided by the House of Orange, the colour yellow or rather orange had become the colour of the Protestant independence movement of the northern provinces. Today it remains the national colour and makes the Dutch sports teams easily identifiable.

Van den Keere would undoubtedly have subscribed to the political interpretation accorded to his 'defiant' Lion and the symbolic role it played in the development of the new nation's consciousness. The Lion's tail curling out into the North Sea may have been simple artistic licence when drawn but it proved appropriate, even prophetic. Some 70 years after the tail was drawn the great grandson of one of these 'rebels' was to cross this sea to become King William III of England (1689–1702).

Antwerp

Antwerp's favourable position as a sheltered port with easy access to inland trading routes can be clearly seen on this extract. This location has proved both an advantage and a handicap. When this map was produced it was emerging as a prosperous city, but only 30 years earlier it was a very different situation. Its strategic position placed it at the centre of Dutch attempts to free themselves from Spanish domination. In 1584 the inhabitants sought to use its low-lying situation to their advantage when, desperate to prevent the Spanish army advancing, they opened the sluice gates and flooded the surrounding lowland (polders). However, the Duke of Parma built a great bridge of boats right across the River Scheldt and cut off the city's lifeline to the sea. Its fate was sealed. Its fall drew England into the conflict and led to the Spanish Armada. While its northern neighbour, Amsterdam, would emerge as the major port of the Dutch Empire as it developed during the next century, Antwerp rebuilt its earlier cultural reputation that continues through to today.

Luxembourg

The front paw of the Lion consists of the province of Luxembourg. While distorted somewhat to fit the limb its position as being surrounded by other countries is clear and this remains so today. Despite its glorious past it was simply one of the 17 provinces when this map was drawn. In fact Luxembourg had been a Duchy since the fourteenth century but, due to its vulnerable location, had been caught up in European power politics for the next 400 years. As part of the Spanish Netherlands it, too, was drawn into the conflicts contemporary to this map. It remained formally attached to the Netherlands until 1839 when it was internationally recognized as an independent state.

While not the richest of the provinces featured on this map it was a prosperous agricultural one. A brief look at the list of provinces in the cartouche box reveals that it contained 23 towns and over 1,100 villages which made it one of the most densely inhabited.

Three Couples

The defiant Lion is not the only political message evident here. There would also seem to be the intention to convey such a message through the way in which the three couples stroll on the same grassy bank on which he crouches. Each pair is engaged in animated conversation but all appear to feel safe in the shadow of the Lion. They seem to suggest that the emergent nationalism presented by the Lion is an inclusive one. The costumes and accompanying notes identify each couple as being (from right to left) from Holland, Belgium and Friesland but there are significant differences. The fashionable sophistication of the Belgian duo contrasts with the pike-bearing Hollander and partner. The plainly dressed and smaller Frieslanders are given a less prominent position. While the message may appear somewhat subliminal, van den Keere would have made the decisions on these images in a considered manner.

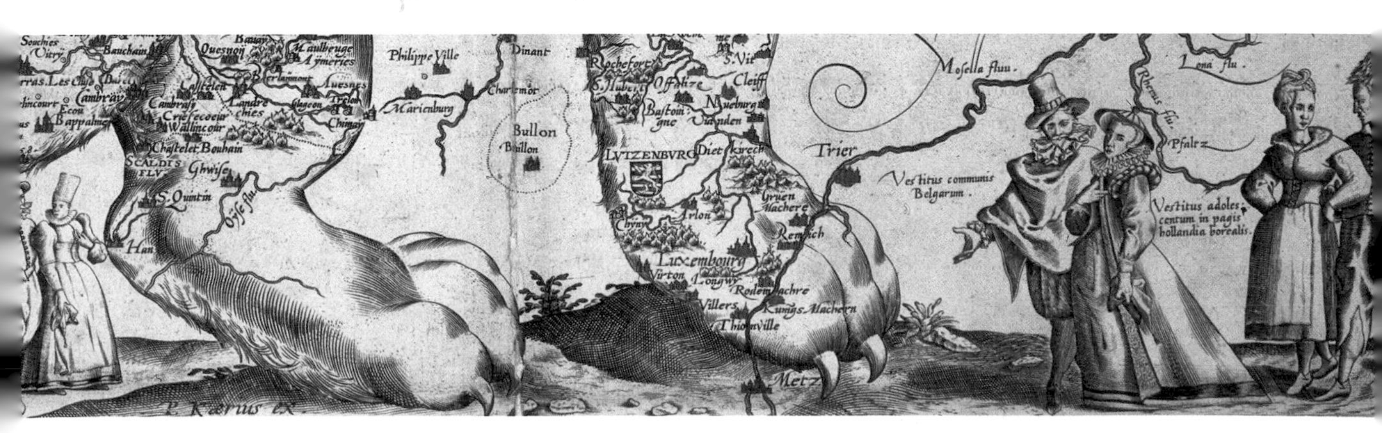

Septemtrion
NUEVO
MEXICO
CANADA
OU
NOUVE
APACHES
DE NAVAIO
APACHES DE
XILA.
NUEVA GRANADA
Taosii
Tami
APACHES.
Tigues
Hubates
Nueva Mexico
VAQUEROS
Zuni als Cibola
Acoma
Amejes
Zaguato
Moqui
Cia
Queres
Peici
Tebes
Tompires
APACHES DE PERILLO
Piri
Socorro
Gorretes
Manses
MARATA REGNUM
Hanes
Sumes
Tomites
QUIVIRA
Tepoanes
Tarrahumares
Iumanes
Tobofes
Passaguates
Iapies
Aixai
Xabotai
Astetlan
Ometlan
Conches
CINALOA
R. del Norte
Rey Coromedo
Lago de Oro
R. de Tecon
R. de Coral
MAR
VERMEIO
CALIFORNIE
ISLE
Agubela de Cato
Talaago
C. Blanco
C. de S. Sebastian
C. de Mendocino
Pto de los Reyes
Pto de Monte Rey
Pto de Francisco Draco
Pto de Carinda
P. de la Conception
Canal de S. Barbara
B. de la Conversion
P. de S. Diego
I. S. Cathalina
B. de Todos los Santos
S. Clement
I. S. Martin
I. de Parraros
I. S. Marco
I. de Ceintas
I. de la Carre
Sierra Pintado
R. de S. Christoval
Punta de S. Apollinar
Pto de la Magdalena
P. de la Marque
P. de Cenou
B. Bernabe
C. de S. Lucas
Las Tres Marias
I. de Gigante
S. Miguel
las Playas
Pto de S. Clara
S. Francisco
MER
DE
SUD OU
PACIFICQUE.
Ligne sous le Tropicque du Cancer.
CULIA
NUEVA BISCAYA
LOS ZACATECAS
S. Iuan
S. Barbara
Durango
NOUVELLE ESPAGNE
GUADALAIARA
Compostella
TERLICHI CHIMECHI
Occident
Midi
Mexico

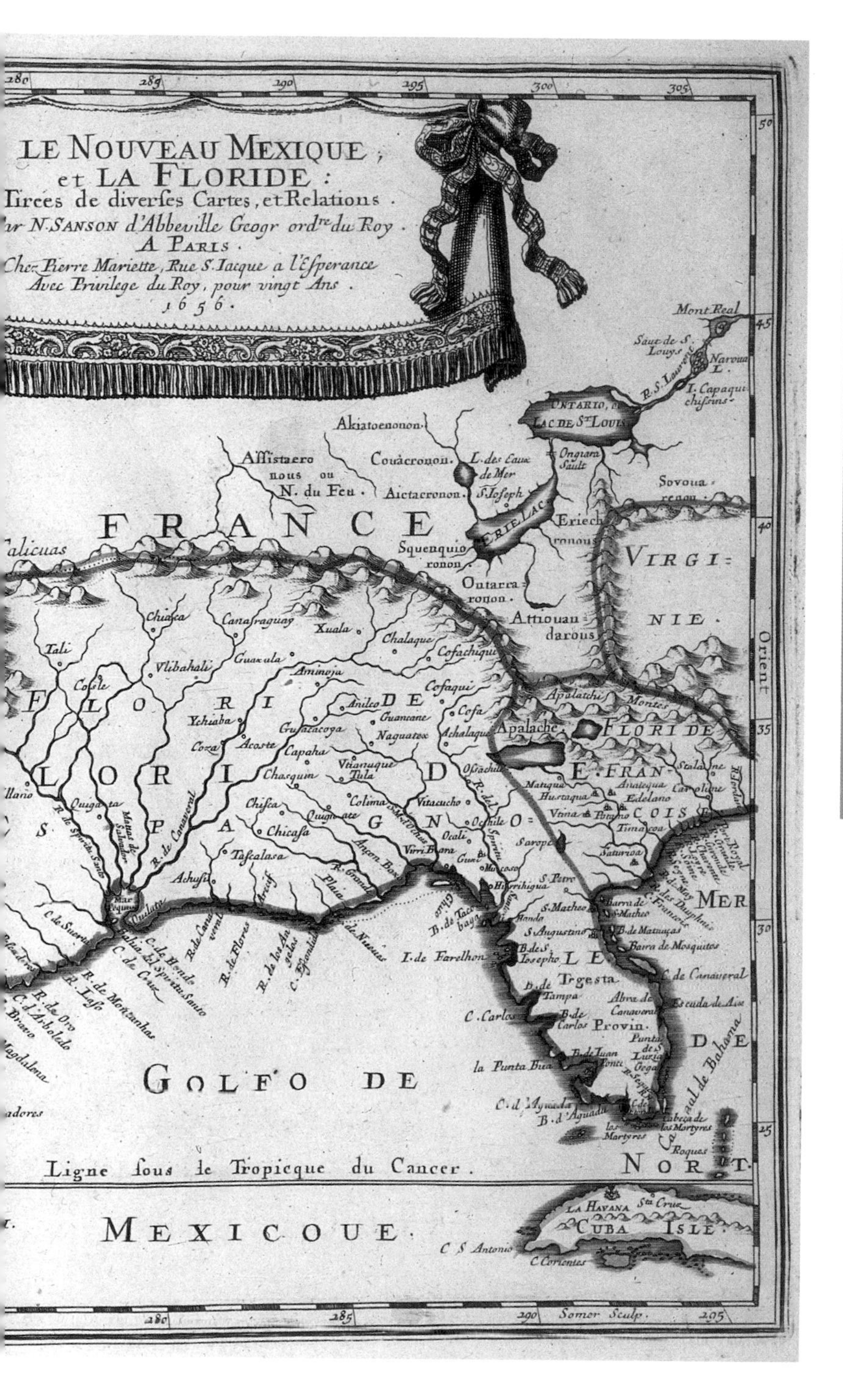

A Novel Idea

The Island of California

By Nicholas Sanson, 1656

A Novel Idea
The Island of California

Over the centuries the labour of the cartographer has led to an improved understanding and knowledge of the world. But at times even the best have got it wrong, which has lead to misconceptions and confusion. The fact that these misconceptions often continued long after maps showing the correct information had been published simply confirms the power a map can exert. This map gave credibility to one of the most famous myths of the 'age of exploration'. Namely, that California was an island and not linked to mainland America. The map was created in 1656 by Nicholas Sanson (1600–1667), one of the great French mapmakers, who, as the cartouche shows, worked from Abbeville, Paris.

Sanson, like other cartographers of the period, was heavily reliant on the information brought back by the captains who took their ships beyond the limits of existing maps and charts. It was common practice for mapmakers to leave large areas of their maps blank or marked 'terra incognita' if they had no details to hand, but new facts were zealously sought and one source was the maps of one's rivals. Plagiarism was rife and there were no rules on copyright. When the island of California first appeared on a map it was rapidly copied without any attempt to verify its validity.

Astonishingly, the myth of 'the island of California' seems to have originated in a novel. Some time around 1500 a book was published in Spain entitled *The Exploits of the Spaniard*, written by Garci de Montalvo. It was an adventure story and one of the characters encountered was Calafia, a queen who ruled over an island kingdom called California, peopled only by beautiful 'Amazon like' black women. It was a 'rip roaring yarn' of a certain genre and one that was apparently enjoyed on the voyages made by Spanish ships to South America during this period. When they ventured further north along the West Coast into the Gulf of California and found land on either side and only water ahead they assumed an island and named it after the one in Montalvo's novel.

However the story doesn't end there. A map showing the newly discovered island was drawn by a certain Father Ascension, a Jesuit priest, and sent back to Spain. On route the ship was attacked and captured by the Dutch and the map was sent to Amsterdam. What happened to it, or the information it contained, is not known but in 1625 the first map to actually show an offshore island named California was published in England. Where, and how Henry Briggs got his information is not known but this lesser known mapmaker must have been delighted with his exclusive. Not wishing to be left behind, more illustrious cartographers such as John Speed, Nicolaas Visscher and Jan Jansson added this new information to their own maps, which strengthened and popularized a myth that proved incredibly durable. By 1705 another Spanish missionary, Father Eusebio Kino, became the first European to walk across on to the top of the peninsula disproving its island status. However, his resulting map was widely dismissed in Europe and the leading mapmaker, Herman Moll, even recounted conversations he had had with British ship captains 'who had sailed all around the island'! As late as 1770 'island' maps were still being produced.

Cartouche from the map provides details about the mapmaker, Nicholas Sanson.

What needs to be acknowledged, of course, is this was simply a European perspective. The indigenous peoples would have established the reality generations earlier but, if they did record it in map-form, these have long since been lost. The 'European perspective' is in evidence right across the map with 'Nouvel le France', 'Floride Espagno' and 'Nueva Granada' being superimposed over large areas with colonial indifference to the native population. The coloured lines marking the accepted, but often disputed, divisions between the European powers would have been drawn thousands of miles away in Europe by people who had not, nor ever intended to visit this part of the world.

While the 'Californie Isle' is perhaps the most fascinating feature of this map today there are others to be found by closer scrutiny. Some features were accurate while others proved to be as mythical as the island that never was.

El Dorado

The existence of a city of gold, or El Dorado, was regarded as fact rather than fiction in Europe during the sixteenth and seventeenth centuries. Men such as Walter Raleigh and Percy Fawcett from England and Cabeza de Vaca, Francisco Pizarro and Francisco Coronado

from Spain undertook costly expeditions in the area that is now Mexico and the southwestern states of the USA. All proved futile and costly in terms of lives, money and reputations, yet the failure of one simply spurred on others. Two locations of 'the city of gold' are shown on this map. Each is plotted in an identical manner to all the other places 'known' to exist, such was their 'factual' status. Cibola was believed to be seven golden cities, while Quivira was not only where the gold mines were but also where Montezuma the Aztec King had sent all his gold to be hidden in its underground caves.

Unfortunately, the Spanish translation of the local word *cibola* as 'gold' was incorrect for its actual meaning was 'buffalo'. There is a parallel, for the buffalo was certainly one of the most valuable things to those living in this part of the world. As for Quivira? Its fabled caves and passages would no doubt have been in the mountain range in whose shadow it stands. Unfortunately, there is no such mountain range.

A Mythic Mountain Range

The tentative state of what was known about this part of North America in the 1600s is clearly illustrated in this extract. Sanson's map is one of the first to name and locate 'Erie Lac' and to show it joined to Lake Ontario, as Lac de St Louis has become known. However, the large lake shown in the mountain range to the south and marked as 'Apalache' has never existed, although the name Appalachian in relation to mountain range has survived. The north–south range shown here is quite accurate but the main range running off to the west and across the centre of the map does not exist. Slowly over the next 200 years an accurate picture of the physical geography of this region would be compiled.

PARADISE, OR THE GARDI
With the Countries circumjacent Inhabited by the
To the Reverend Father in God HENRY Lord Bishop of London. This Mapp Is humbly Dedicated by J. Moxon.
Capadocia
ARMENIA MINOR
or
Armenia the Less.
ARMENIA or Armenia the Great.
The Hills of Taurus
Comagena
PADDAN-ARAM.
Haran the City of Nahor, sometimes called Aram and also Charan
CILICIA.
MINOR.
THE PAMPHILIAN SEA.
Ciprus
THE SIRIAN SEA.
The Sea of Issus
SYRIA.
Phoenicia
Aram
Palmyrena.
Dedan
Cathany
THE MEDITERRANEAN SEA called in the Bible THE GREAT SEA.
Jonas flies from the Lord
THE EGIPTIAN SEA.
Japho or Joppe from whence Jonas set out to Sea.
the Land of Hus
Canaanites
Gergesites
Hevites
Amorites
THE
OF CUS,
Saccea
Agreni
Nabathea in Hebrew Nabajoth
The way that Jacob with his Family and Cattle travelled when he fled from his Father in Law Laban to his own Countrie
AMMONITES
MEDIANITES
ARABIA THE STONY
ISMAELITES
called THE DES
THE LAND OF CHAM also THE LAND OF GOSEN
MITS OF EGYPT
Grand Cair or Babilon of Egypt
Desert of Pharan
Desert of Shur
Desert Zin or Kadesh
Edomites
LAND OF MADIAN
The Red Sea.
ARABIA THE HAPPY that is the Kingdom of SHEBA from whence the Queen of Sheba came to hear the Wisdom of Salomon and presented him with Gold, Jewels, etc.
The Mountains of Arabia
ISMAE

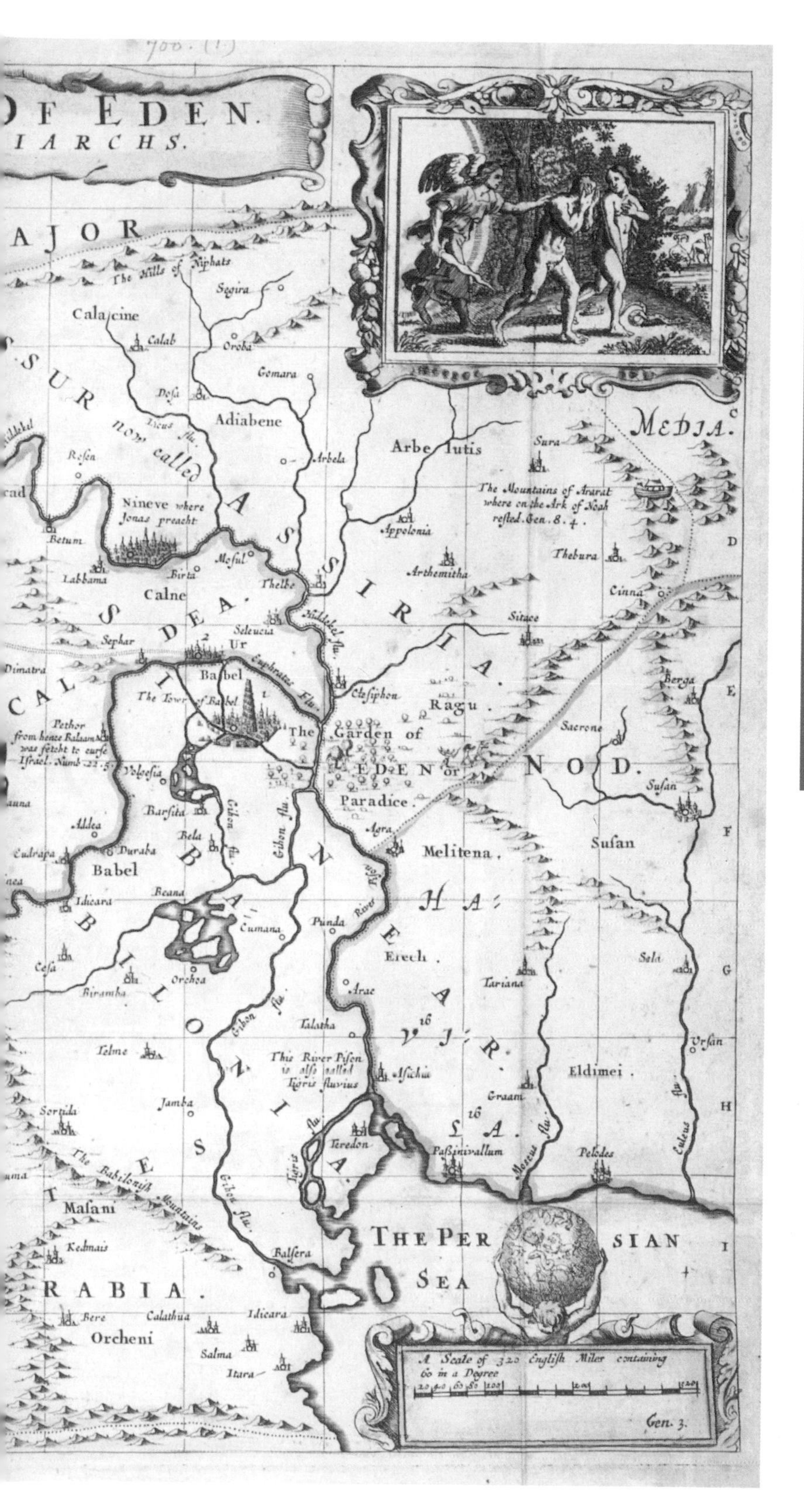

Plotting Paradise

The Garden of Eden

By Joseph Moxon, 1695

Plotting Paradise

The Garden of Eden

Even if one is not religious, the use of the phrase 'Garden of Eden' conjures up images of unspoilt beauty and natural simplicity. Whether God or nature is viewed as the creator, 'Eden' continues to suggest a world as it was intended to be – an earthly Paradise. Today, there are different interpretations of the Bible and its stories but it needs to be remembered that until quite recently it was regarded as *literally* true by all whose religious beliefs it informed, and there are a significant number who continue to regard it in this way today. From this literal position the locations of the places it mentions are identifiable and can therefore be mapped. This map confidently addresses the location of Paradise and its relation to the surrounding area.

The essential problem for the mapmaker who created this map was that it no longer existed. Unlike many of the other features shown, it could not be visited to verify its position for, as the Book of Genesis recounts, it had not survived the expulsion of Adam and Eve. While problematic, this is a challenge mapmakers, historians and archaeologists have always faced. The existence and location of cities such as Troy and El Dorado, landmarks such as the Pillars of Hercules and even whole civilizations such as Atlantis, continue to fascinate each new generation. The search for their actual location has involved returning to original sources and looking for clues and evidence. In the case of the Garden of Eden this takes the researcher back to the Bible where in Genesis (2:10–14) one reads:

> And a river went out of Eden to water the garden; and from thence it was parted, and became into four heads. The name of the first is Pison: that is it which compasseth the whole land of Havilah, where there is gold; And the gold of that land is good: there is bdellium and the onyx stone. And the name of the second river is Gihon: the same is it that compasseth the whole land of Ethiopia. And the name of the third river is Hiddekel: that is it which goeth toward the east of Assyria. And the fourth river is Euphrates...

All these names appear on this map but while the River Euphrates is easily identifiable today the Pison, Gihon and Hiddekel have confounded scholars for over a thousand years and continue to do so. The classical identification of them with the Nile, Ganges and Tigris proved impossible to sustain, at least in relation to the first two, once a more detailed knowledge of topography and distance developed. The reference to Ethiopia proved particularly problematic because of its relative position and distance to the more recognizable locations. The hunt for an 'actual' as opposed to a 'metaphorical' Garden of Eden continues into the twenty-first century and the internet abounds with reports of archaeological digs, of satellite photography, of newly discovered dried river beds, of rising sea levels and other hypotheses that claim to have *finally* located this ancient Paradise.

'The Creation' from a late fifteenth-century prayerbook. It shows the Creator above a tranquil landscape with animals, including a horse, a lion and a camel.

This map is in the tradition of the Psalter map featured in Chapter 1 in that it seeks to locate key events and places mentioned in the Bible narrative. The final resting place of Noah's Ark, the Kingdom of the Queen of Sheba, and the Tower of Babel, among others, are all identified by finely engraved notes. The mapmaker helpfully relates the modern places to their earlier biblical names noting 'Assur now called Assiria' and 'The Mediteranean Sea called in the Bible the Great Sea'.

The map dates from around 1695 and was originally produced in London by Joseph Moxon (1627–1691). His map was, in effect, a copy of an earlier Dutch one (1671) by Nicolaas Visscher who would himself have called upon contemporary and earlier maps as he created his own. Moxon translated the engraved text from Dutch to English which was something in which he specialized.

The 'painting' chosen for inclusion in the top right corner is there for commercial reasons, its function being similar to that of a 'titillating' picture on the cover of a popular fiction book today hoping to attract the attention of a potential buyer and thus lead to a sale.

Out!

Besides locating the Garden the map also tells the story of the events that occurred there through two small and separate pictures, almost in what we would call cartoon style today. In the Garden itself, the slightly risqué drawing of an aroused Adam and a seemingly eager Eve would surely have amused its late seventeenth-century audience. It is inconceivable that such

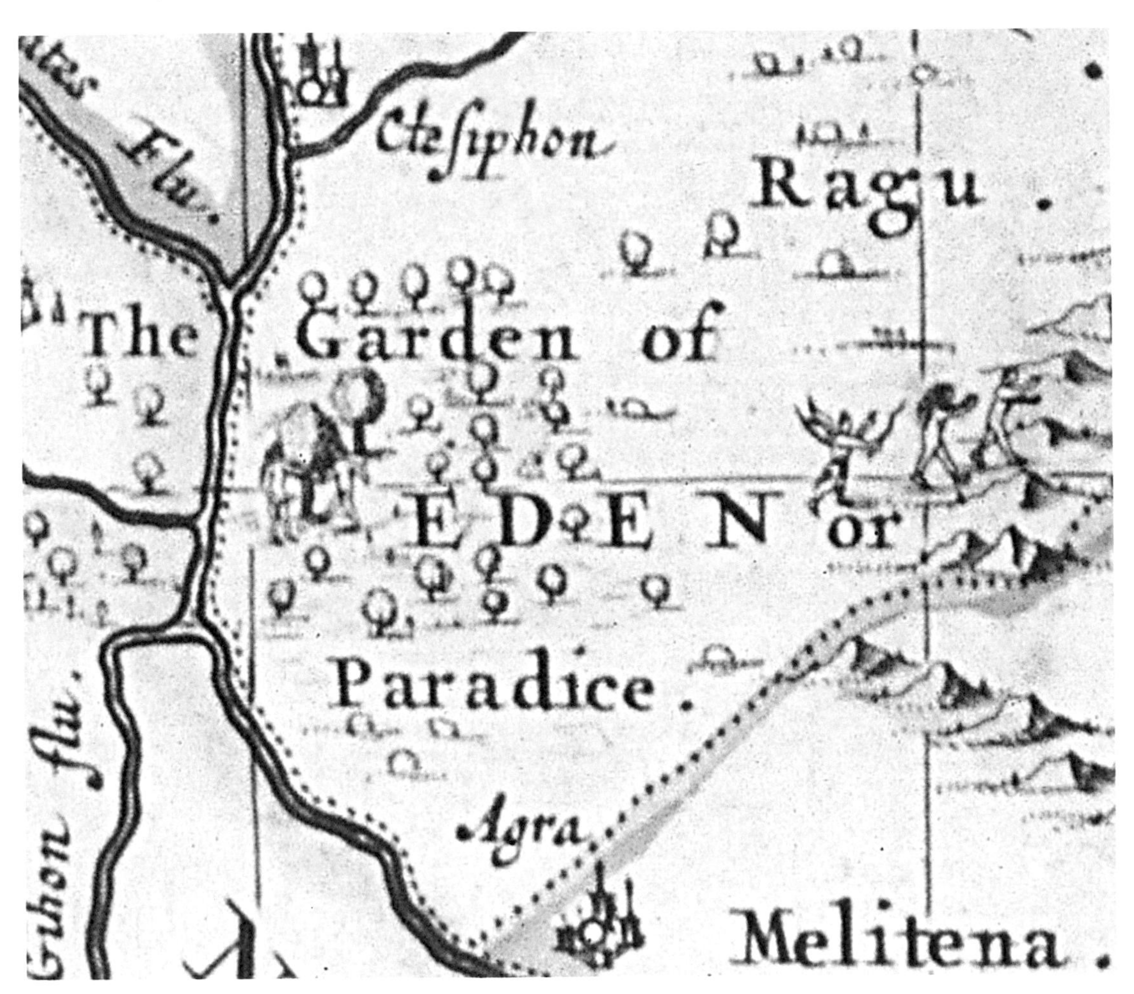

an image would have been permitted some 40 years earlier when Britain was under the control of Cromwell and the Puritans.

The second part of the story can be seen to the right of the Garden as the consequences of Adam and Eve's affections are made apparent. The armed, winged Angel is seen driving them into the mountains and towards the Land of Nod. While the snake is in the painting at the top of the map it does not seem to be in evidence here.

Seeking to Escape

Jonah and his exploits with the whale have always been a popular biblical story. As Moxon shows, its actual location can be accurately plotted on a map and he records where the events took place. The port of 'Japho or Joppe from whence Jonas set out to sea' is clearly

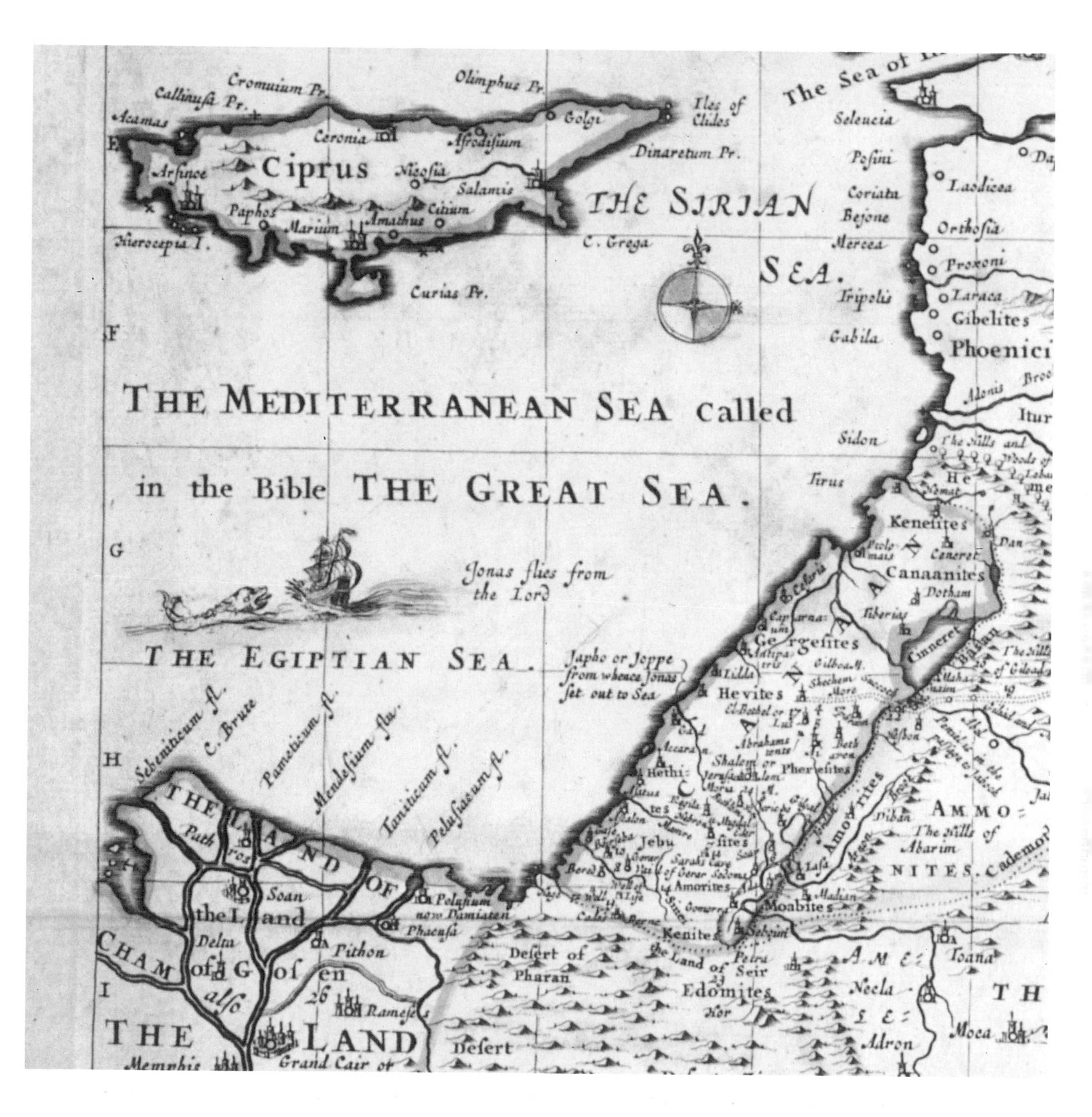

marked on the 'Great Sea'. The large and fierce-looking fish is shown pursuing a ship in full sail as 'Jonas flies from the Lord'. As the story recounts he was unsuccessful in his attempt to escape and spent three days inside the whale's stomach. It is interesting to note that in the Old Testament Book of Jonah (1:17) it is a 'great fish' that features in the story and only when Matthew (12:40) refers to the story in the New Testament is it identified as 'a whale'. Whatever, the picture of the map captures the essence of a story that would have been very familiar to its audience.

It is interesting to see that the ship reflects the engraver's knowledge of seventeenth-century European ship design rather than that of the biblical period.

Sketch of the Coasts of
Australia.
and of the supposed Entrance of
The Great River:
also a Diagram of the
Surveying Stations,
in the Interior.
Drawn for the
Friend of Australia.
1827.
AUSTRALINDIA
WEST AUSTRALIA
ANGLICANIA
THE DELTA of AUSTRALIA
The Great River or the Desired Blessing
Route recommended for
Commencement of very elevated Land supposed by some
Perth
Fremantle
Geographe Bay
C. Naturaliste
Blackwood
SUSSEX
King George 3d Sound
Archipelago of the Recherche
Nuyt's Land
Spencer's Gulf
Kangaroo Id.
St. Vincent's Gulf
Investigator's Strait
Arnhem's Land
De Witt's Land
Dampier's Land
Endracht's Land
Edel's Land
TIMOR
C. Londonderry
C. Arnhem
Lat 23
Long 132°
Depôt
Hammet Island

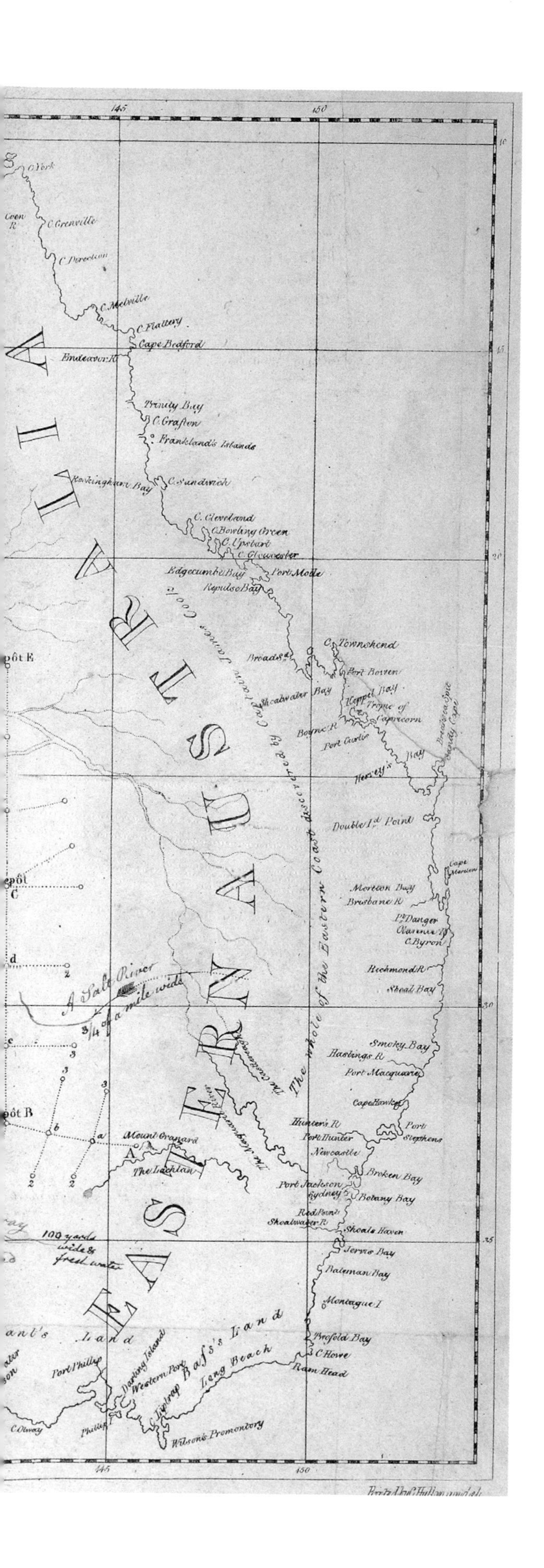

A Logical Necessity

The Great River of Australia

By T. J. Maslen, 1827

A Logical Necessity

The Great River of Australia

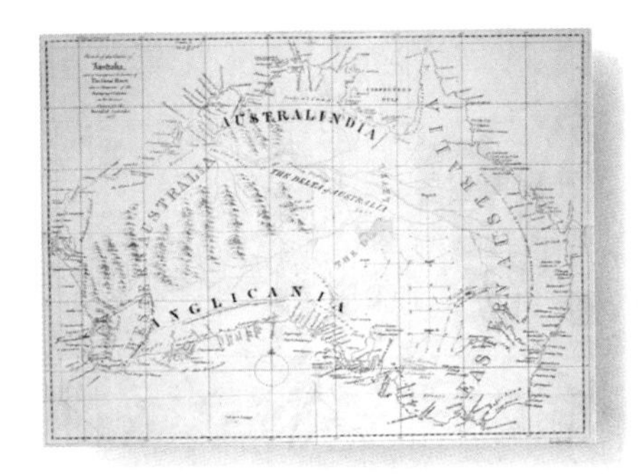

Whether the desire to explore is a natural human instinct or is motivated by more mundane motives continues to be debated. Whatever the motivation of the captains and crews on the tiny sailing ships that set off into uncharted oceans, the cumulative effect of their many voyages was to lead to the map of the world being completed. The role of the cartographer has been to collect and collate this knowledge and to use his skills to represent it visually so that it can be shared. Until relatively recently cartographers were dependent on others to provide the information they worked with. As water travel was the main, and indeed in most parts of the world the only form of long distance travel, it is not surprising that they had more information about coasts and rivers than anything else. The maps they drew reflected this and the European maps of Asia, Africa, the Americas and Australasia were initially little more than outline maps. Once these outlines were established the challenge to fill in the detail began. There was great kudos in having a new detail on your map. On occasions this led to some fanciful and particularly creative interpretations of the information available, which produced bizarre results. This map of Australia is a fascinating result of this process.

While its overall shape is much as we recognize it today it is the interior that is intriguing. Its title states that it shows 'the supposed entrance of The Great River'. As can be seen, this river enters the Indian Ocean towards the top left corner of the map. It is only when one begins to trace its course inland that its magnitude becomes evident. Its first meander takes it through a gap in a broad mountain range before continuing across the centre of the continent. One of its tributaries flows from a giant central lake while others rise in what we now know as the Great Dividing Range near the eastern coast. One actually appears to rejoin the sea at Broken Bay just to the north of Sydney. Such a river would have a length in excess of 3,000 miles (4,830 kilometres). Today we know that there is no such river and, because of the information we

Aborigines have always had a deep spiritual bond with the land and an understanding of its topography and river systems.

now possess about the topography of central and Western Australia, we also know that it could never have existed as depicted here.

It needs to be remembered that at the time it was drawn the European knowledge of Australia was incredibly limited. It was barely 50 years since Cook had established that it was an island and less than 40 years since The First Fleet had landed. The creator of this map, T. J. Maslen, was just one of the many Europeans who was fascinated by the sheer vastness of the continent. His map was not simply a product of his imagination: he had clearly read the accounts of those who had actually been to Australia. Joseph Banks had sailed with Cook on the voyage that first charted the eastern coast and was a leading figure in the scientific community. In 1798 he had concluded that as Australia was as large as Europe it must have vast navigable rivers that went into the heart of the continent. It was an opinion to be respected. When Matthew Flinders, captain of the first ship to complete a circumnavigation of Australia (1801–03) pondered that there might be an inland sea of Mediterranean proportions, interested people again took note. Maslen had worked in India but had never visited Australia and the map he created over 10,000 miles (16,100 kilometres) away in England was a combination of the known and the hypothetical.

While his title is tentative in one respect – 'the supposed entrance of the Great River' – its actual existence is presented as fact. The belief in a great river and/or the inland sea continued for many years to come. When Charles Napier Sturt, perhaps Australia's greatest explorer, set off into the interior nearly 20 years after this map was published, he took a boat with him as an essential piece of the expedition's equipment.

The modern map of Australia shows that there is a river that enters the ocean at the approximate location shown on Maslen's map. This river does have the same initial wide sweeping bend as it heads inland. However, the Fitzroy River is a much more modest one than Maslen believed and it does not provide the desired route into the interior.

'The Great River' was to be relegated to join the not inconsiderable collection of fascinating but fantastical features that have appeared on maps at different times in and in different locations around the world.

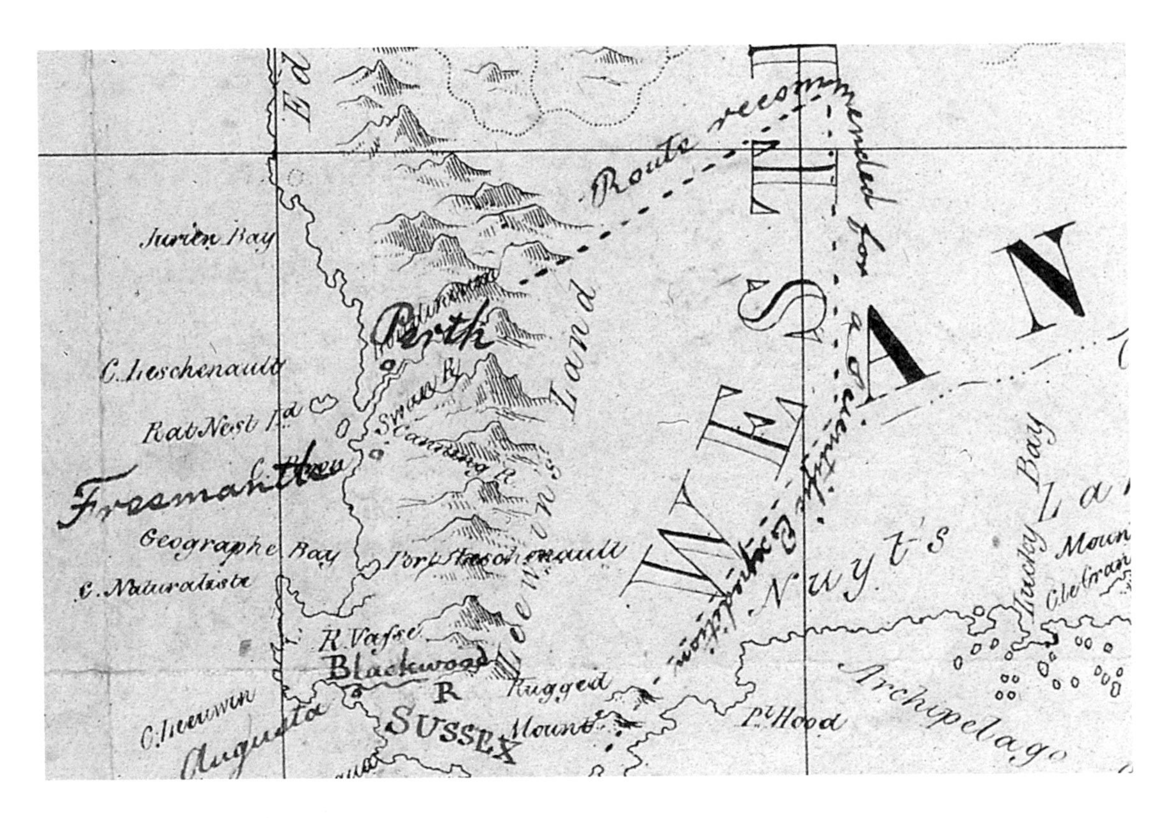

Interesting Annotations

This map has a number of hand-written annotations that add to its interest. When it was published in 1827 there were no Europeans living in this area for this had been the land of the Nyoongar tribe for thousands of years. In the late seventeenth century Dutch sailors had stopped here briefly to replenish their supplies and named the river the Swan after its distinctive natural inhabitants. They also named the land New Holland and for over a century maps of the western side of Australia carried this name. It was not until 1829 that the British formed a permanent settlement here. After leading a surveying expedition, Captain James Stirling asked for, and was granted, permission to set up a colony. In classic colonial fashion he had sent a party on ahead to formally claim the land for the British Crown. The leader of this party was Captain Charles Howe Fremantle (just one 'e' in the surname), and when a coastal port was developed to serve the new colony it was named after him. Perth itself was originally the

Swan River Colony and its name was changed by the early settlers to reflect their Scottish links.

Who the dotted line on the map was drawn for and whether and when the 'route recommended for a scientific expedition' was ever undertaken remains uncertain. Its projected route would have taken it into the Australian gold fields whose discovery led to the gold rush and massive population influx only 50 or so years after this map was first published.

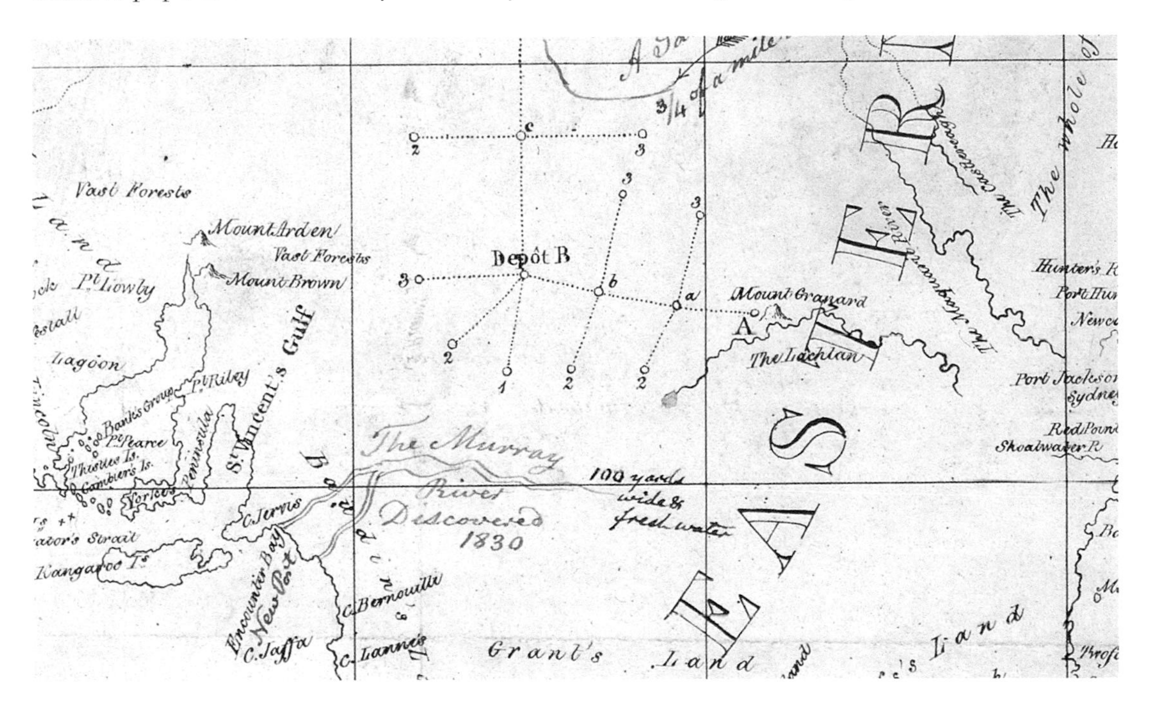

Imaginary Depots?

This extract captures the cumulative manner in which information was built up as this part of the world was first mapped by the new arrivals from Europe. The lower reaches of the Murray River have been drawn on and annotated 'Discovered 1830', but the further additional notes on its width and that it was 'fresh water' would have been more vital to those in this region. The top of this extract has another river marked but not named and annotated 'A sale River 3/4 of a mile wide'. This would eventually be named the River Darling and its course into the Murray plotted to form the major waterway of the continent. The concept of 'discovery' is of course a relative one here for the Aborigines would certainly have known of its existence.

The 'depots' carefully drawn and labelled on the main map (and included in this extract) are very confusing, because although the title states the map is 'also a diagram of the surveying stations in the interior' none of this had occurred by 1827! It would seem that Maslen was superimposing his experience from surveying in India to the vast empty areas of the interior of Australia. It was not until after this map was published that Charles Napier Sturt began to explore this region and not until 1860 that Robert O'Hara Burke and William John Wills reached the Gulf of Carpentaria (shown as 'Carpenters Gulf') from the southern coast. It must have looked so much easier as Maslen drew his map at home in England, but the reality was much different. Burke and Wills died on their way back and Napier Sturt's health was seriously damaged by his explorations.

UNDERGROUND TRAIN IN STATION

Exploring Further

The intention of this book has been to explore the stories associated with the maps included and to illustrate how a map can open up a window on a particular period in history. As I conducted my research I invariably found out more than I was able to include in the space available here. It is likely, indeed it is hoped, that in relation to at least one of the maps the reader has been left wanting to 'know more'.

The strategies and references given below should enable anyone who now desires to continue the exploration of the maps included here or to open up their own areas of interest. Invariably the internet features prominently as a starting point for these pursuits. As is the case in so many areas of research it has literally revolutionized the ordinary person's access to maps held across the world and many of these can now be viewed on screen. This ease of access has never been greater although it is not without its limitations. Perhaps the most significant of these relates to the size of the map because the computer images have often been increased or more often decreased in size to fit the screen and this can create an impression of uniformity that is misleading. Viewing works of art using the computer poses the same problem. However, while an awareness of this needs to be acknowledged the advantages far outweigh this 'problem'. One such advantage is that the technology allows the user to 'zoom in' on a particular area or aspect of interest, which allows the detail to be examined in a way that has often not been possible when viewing the originals at an exhibition or looking at reproductions in traditional book format.

However, it is another feature that I find particularly exciting and which is likely to be used more and more in the future. This consists of the ability to link text with the map itself in an interactive way. This can be seen to excellent effect with regard to the John Smith map featured in Chapter 2. On the internet version of his map the routes of his exploratory voyages into Chesapeake Bay have been plotted. As one moves along the route the text from his journal relating to the features and locations appear on screen. This is a wonderful combination of two original historical sources in a mutually enhancing manner and the creators of this site <www.mariner.org/chesapeakebay/colonial/col009.html> are to be congratulated. The possibilities for undertaking something similar with some of the other maps featured in this book are very exciting. Imagination, commitment, time and energy are the ingredients that need to be added to the technology that is now available.

There is, I would argue, no substitute for actually standing in front of an original, and thematic exhibitions of maps will continue to be prove popular. But technology is also changing the traditional manner of these. Recent map exhibitions at The British Library, including 'The Lie of the Land' of 2001–2002, allowed visitors to view the originals while its website enabled these to be explored further in an interactive way both during and after the exhibition had closed. This model is likely to become increasingly common and is a further step in allowing more and more people to enjoy their national and international map heritage.

The internet has also become the means through which to find out what is going on in the world of maps. Information about international, national and local exhibitions is now disseminated through this medium as is information about new books, maps for sale and any new developments in the field.

The following are therefore no more than possible starting points for further exploration and the reader is advised to use their favourite search engine (e.g. Google) to explore an area, a person or a period they would like to know more about.

A Good Place to Start

<http://www.maphistory.info/index.html>

This is an excellent place to begin as this site's structure allows the user to identify and then follow her/his own lines of enquiry or interest with ease. The manner in which it links with other sites is impressive. It is maintained by Tony Campbell who was Map Librarian at the British Library until his retirement.

An Interesting Website

For the real underground map, as mentioned in Chapter 3:
<http://tube.tfl.gov.uk/content/tubemap/realunderground/realunderground.html>

Places to Visit

While the national museums invariably hold the largest collections of maps, regional museums and specific interest museums (regimental, local history, etc.) often hold unique examples relating to their sphere. Knowledgeable people who are more than willing to share their enthusiasm usually staff them. Again the internet will enable the reader to initially find out what exists and where and then check out important details such as opening hours, etc.

The following all have maps on display on a permanent basis and their web sites are given so that the details of any current/special exhibitions can be ascertained.

Bodlean Library
<http://www.bodley.ox.ac.uk/guides/maps/maproom.html>

National Maritime Museum
<http://www.nmm.ac.uk/>

British Library
<http://www.bl.uk/>

National Library of Scotland
<http://www.nls.uk/pont/>

To Read

There are hundreds and hundreds of books written on different aspects of maps and cartography. These range from hefty academic tomes to small and inexpensive 'Discover Maps' type booklets sold in non-specialist book outlets. As ever, it depends what you are looking for.

For an authorative overview of the history of maps and their creators John Noble Wilford's *The Mapmakers: The Story of the Great Pioneers in Cartography – From Antiquity to the Space Age* (Pimlico, 2002) is an excellent read.

For anyone considering collecting maps Jonathan Potter's *Collecting Antique Maps: An Introduction to the History of Cartography* (Jonathan Potter, 2002) is both informative and accessible.

For the reader who may be interested in reading novels that have maps as a central theme there have been a number of excellent titles published in recent years.

The following are successful in being both good literature and historically accurate.

Cadbury, D., *The Dinosaur Hunters* (Fourth Estate, 2001)
Harvey, M., *The Island of Lost Maps: A Story of Cartographic Crime* (Phoenix, 2002)
Moxham, R., *The Great Hedge of India* (Constable, 2001)
Sobel, D., *Longitude* (Fourth Estate, 1998)
Winchester. S., *The Map That Changed the World: The Tale of William Smith and the Birth of a Science* (Penguin, 2002)

The following will prove of interest for those seeking a more in-depth treatment.

Aczel, A., *The Riddle of the Compass* (Harcourt Brace, 2001)
Barber, P., *Tales from the Map Room* (BBC Books, 1993)
Black, J., *Maps and History: Constructing Images of the Past* (Yale University Press, 2000)
Keay, J., *The Great Arc: The Dramatic Tale of How India was Mapped and Everest was Named* (Harper Collins, 2000)
Short, J. R., *The World Through Maps: A History of Cartography* (Firefly Books, 2003)
Thrower, N. J., *Maps and Civilization: Cartography in Culture and Society* (Press, 1999)
Whitfield, P., *New Found Lands: Maps in the History of Exploration* (Routledge, 1996)

Map and Picture Sources

Front Matter

Jacket The National Archives (TNA): PRO FO 925/4111; **2** Lauros/Giraudon/Bridgeman Art Library; **8** Private collection/Bridgeman Art Library **10** British Library Eger.1894 **12** TNA: PRO MPI 1/168 **14** TNA: PRO MPF 1/1 **15** The Art Archive/Museo della Civilta Romana Rome/Dagli Orti

The Early Mapmakers

18 British Library Add.39943/Bridgeman Art Library **20** British Library IC.9304 **23** akg-images/Erich Lessing **26** akg-images **29** Topham Picturepoint **32** British Library Add.28681 **35** akg-images/Juergen Sorges **38** British Library Cott.Claud.D.Vi **41** British Library Royal.18.D.II **44** TNA: PRO E 164/25 **47** Chertsey Museum

The Quest for Riches

53 Giraudon/Bridgeman Art Library **54** akg-images **56** Werner Forman Archive **57** Private collection/Bridgeman Art Library **60** British Library Add.5415A **63** Lauros/Giraudon/Bridgeman Art Library **66** TNA: PRO MPG 1/284 **69** *Travels and Works of Captain John Smith* **72** British Library C.4.b.1(59) **74** akg-images **75** Topham Picturepoint **78** British Library 455.a.23 **80** National Maritime Museum **81** TNA: PRO ADM 55/40 **82** British Library 214.c.6/8 **83** British Library 214.c.6/8

Challenging Perceptions

87 Ken Garland/Transport for London **88** British Library K.Top.118.49.b **90** akg-images **91** akg-images **94** British Library 144.e.24 **97** Science Museum **100** The Library of the London School of Economics and Political Science **102** The Library of the London School of Economics and Political Science **103** TNA: PRO 30/69/1663 **106** TNA: PRO CO 956/537A **108** TNA: PRO CO 956/553 **109** TNA: PRO CO 956/426 **112** Transport for London **115** Mary Evans Picture Library

Winning the Day

120 TNA: PRO WO 1/205/2 **122** TNA: PRO MPF 1/1 **125** Topham Picturepoint **128** TNA: PRO WO 78/1006 **131** Topham Picturepoint **134** TNA: PRO FO 925/1833 **137** Peter Newark's Military Pictures **140** TNA: PRO WO 153/114 **143** Topham Picturepoint **146** TNA: PRO AIR 41/24 **149** TNA: PRO AIR 25/792

Fantasy and Fantastical

155 British Library 700/1 **156** British Library C.116.g.2 **159** Private collection/Bridgeman Art Library **162** British Library C.7.c.10 **165** British Library C.22.c.12 **168** British Library C.36.f.6 **171** British Library C.36.f.6 **174** British Library 700/1 **177** British Library Add.18851 **180** TNA: PRO MPG 1/681 **183** Penny Tweedie/Panos **186** Mary Evans Picture Library